The Sun is a Plasma Star

Illustrated Science Exploration by Rolf A. F. Witzsche

© Text Copyright Rolf A. F. Witzsche 2018
all rights reserved

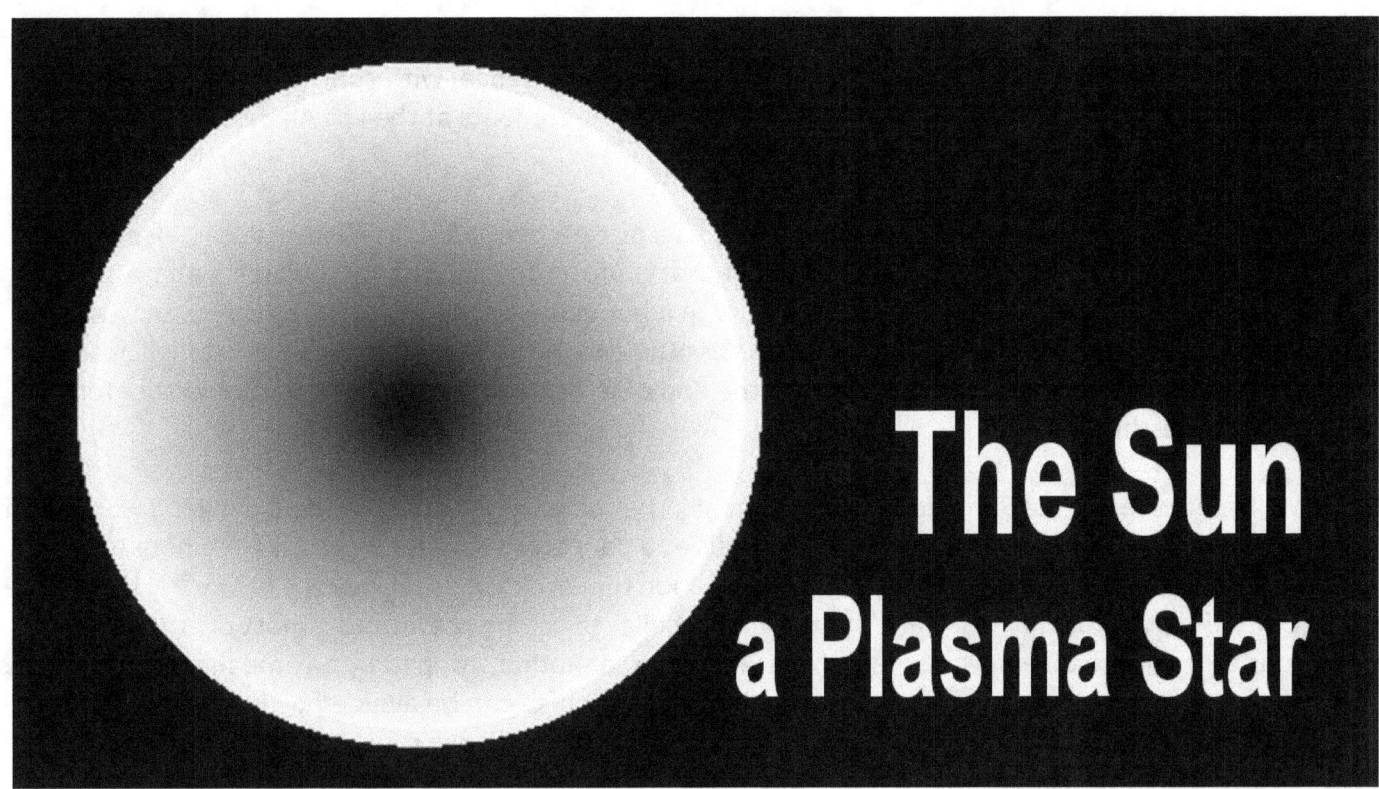

This book contains the transcript with images of the exploration video with the above title:
see: http://www.ice-age-ahead-iaa.ca/

Lead in:
A society that doesn't believe in science,
or only superficially - that believes in
opinions, doctrines, dreams, and myths, -
has nevertheless the capacity to awake.

The Gas Sun theory, the darling of mainstream cosmology by near universal consensus, is so full of holes that a whopping eleven major paradoxes put the Gas Sun theory deep into doubt, and effectively invalidate it.

Paradox #1 - The Sunspot Paradox
Paradox #2 - The Solar-Wind Paradox
Paradox #3 - The Solar-Wind-Acceleration Paradox
Paradox #4 - The Super-Hot-Corona Paradox
Paradox #5 - The Low-Mass-Density Paradox
Paradox #6 - The Giant-Stars Paradox
Paradox #7 - The Solar Cosmic-Rays Paradox
Paradox #8 - The Solar-Cycles Paradox
Paradox #9 - The Sunlight Paradox
Paradox #10 - The Atomic Elements Paradox
Paradox #11 - The Differential Rotation Paradox

The Plasma Sun concept, in contrast, has none of the paradoxes attached, of the Gas Sun theory, but is instead supported by a wide range of measured physical evidence, with contributions by NASA, the European Space Agency, ice coring projects, laboratory experiments at the Los Alamos National Laboratory and by other related experiments.

The experiments and discoveries that present us a picture of our Sun as a Plasma Star, also reveal our Sun to be an immensely variable star, and a star that is presently diminishing towards a phase shift to the next Ice Age in potentially the 2050s. We are in a boundary time-zone towards this event, with numerous fringe effects telling us that we are near the end of the line, and that a climate collapse is in progress that is accelerating, - that promises the end of the current form of agriculture long before the Ice Age phase-shift begins.

What sets the Plasma Sun effects apart from all other dangers - such as political dangers, wars, even economic collapse, which have grown so ominous and dangerous that they can take the house of civilization down to the point of causing the extinction of humanity, as in the case of war - is the simple fact that the political dangers, wars, and economic collapse can be prevented almost overnight if the actions to do so are taken, while the Ice Age Challenge cannot be stopped by any means that we process, with consequences that only very few would survive, but which can be avoided by humanity building itself a technological new world that the now ongoing climate collapse cannot affect.

The point is, that the political existential challenges are all optional, while the cosmic Ice Age Challenge is not. That's the difference. The difference is enormous. It is dangerous. And it is tragic, because the field for the strategic defense of humanity is empty. The most precious gem on Earth, which is humanity, remains presently unprotected and on a path determined by ignorance, poised to commit suicide. Whether humanity will heal itself in time, of its 'smallness', remains yet to be seen. The potential exists for humanity to awake. Here lies our hope to have a future, a bright future by necessity. The Plasma Sun stands at the forefront of this hope as a key element to inspire actions.

Table of Contents

False consensus: The Sun is a sphere of hydrogen gas 10

    The Sunspot Paradox 11

    The dark umbra of the sunspots is theorized 'dark energy' 12

    The Solar-Wind Paradox 13

    The Solar-Wind-Acceleration Paradox 14

    The Super-Hot-Corona Paradox 15

    The Low-Mass-Density Paradox 16

    The Giant-Stars Paradox 17

    This star exists, and it outshines our Sun 340,000-fold 18

    The Solar Cosmic-Rays Paradox 19

    Cosmic-ray fluctuations with the Sun's activity cycles 20

    Synchronism measured by the Moscow Neutron Monitor 21

    The Solar-Cycles Paradox 22

    The Sunlight Paradox 23

    The Sun emits a richly homogenous spectrum 24

    The Atomic Elements Paradox 25

    The Differential Rotation Paradox 26

    Consensus model for the Sun is fundamentally wrong 27

    Differences in Mass Distribution 28

The Plasma Star: A leading-edge concept in astrophysics 29

    A plasma star is a large sphere of plasma 30

    Plasma is the name for subatomic particles 31

    When plasma particles are free 32

    In plasma, the tiny electron is intensely drawn to the big proton 33

- **A large plasma sphere has its least mass density at its core** 34
- **With its electron-rich surface, a large plasma sphere is a natural sun** 35
- **A plasma sphere is a plasma star** 36
- **A plasma star is essentially a largely empty sphere of plasma** 37
- **Sun's energy is generated on its surface** 38
- **A wide emission spectrum** 39
- **The Plasma Sun concept eliminates ALL paradoxes** 40

The Plasma Sun solves Paradox #1 41
- **A plasma sun doesn't actually create energy** 42

Paradox #2, the solar-wind paradox 43
- **Solar wind is only possible by a plasma-fusion process** 44

Paradox #3, the solar-wind acceleration paradox 45
- **The expansion is powered exclusively by the electric force** 46
- **Solar wind likened to a heated kettle boiling off steam** 47
- **For as long as the solar wind flows** 48
- **When the heat is turned down on a boiling kettle** 49
- **The Sun 'cools' down after the solar wind stops** 50

Paradox #4, the super-heated solar corona paradox 51

Paradox #5, the mass-density paradox 52
- **The mass-density difference between the Sun and gas planets** 53

Paradox #6, the large-star paradox 54
- **The star UY Scuti has its ten solar masses spread across its thin plasma shell** 55

Paradox #7, the cosmic-ray paradox 56
- **Solar cosmic-ray flux is generated on the surface of the Sun** 57
- **A profusion of single plasma particles escaping the fusion cells** 58
- **Cosmic-ray flux originates with the Sun** 59
- **That most of Carbon-14 is caused by the Sun** 60

When the solar activity is strong ... 61

That the fluctuating cosmic-ray flux is solar in origin ... 62

A third proxy for solar cosmic-ray flux ... 63

A gas star is cosmic-ray dead ... 64

Paradox #8 of the 11-year solar cycles ... 65

The gas-sun model doesn't support solar cycles ... 66

The solar cycles are not caused by the Sun itself ... 67

Paradox #9 is located in the sunlight ... 68

Homogenous spectrum of colors in the sunlight ... 69

Paradox #10 is the paradox of the profusion of atomic elements ... 70

The plasma Sun synthesizes all atomic elements right on its surface ... 71

Only the plasma Sun can generate the rich spectrum of the sunlight ... 72

Paradox #11 of the differential rotation of the Sun ... 73

A plasma Sun operates by drawing interstellar plasma streams ... 74

When the magnetic fields tangle up ... 75

The plasma stream surrounds the Sun ... 76

An outer ring of 56 plasma filaments with rotating magnetic fields ... 77

The Sun becomes surrounded by rotating magnetic fields ... 78

The Sun's differential rotation proves that the Sun is a plasma star ... 79

Solar operation consumes a portion of the in-flowing plasma ... 80

All the atoms that exist, were synthesized on the surface of a Sun ... 81

The consumption of plasma creates the sink effect that enables plasma to flow ... 82

All the atoms for the planets were synthesized by the Sun ... 83

Whatever portion is not consumed, simply flows on ... 84

That's why the stellar primer fields are complimentary in nature ... 85

The out-flowing stream subsequently gathers up plasma ... 86

The term, 'the Primer Fields' ... 87

Stars are often lined up into long strings of stars in cosmic space 88

When the plasma flow becomes too weak 89

The deep Ice Age resumes that has not been experienced 90

The hibernating plasma Sun has measured evidence 91

The Sun is presently in its interglacial high-powered mode 92

A complete void of solar-wind movement over the poles 93

Two gigantic plasma structures 94

The plasma structures are naturally visible in gamma-ray and x-ray light 95

The similarity in shape and location 96

The entire galaxy as a part of a node-point structure 97

The discovery thereby obsoletes the Big Bang theory 98

The origin of the cosmic plasma streams 99

The theoretical physicist David Bohm 100

Empty space as not at all empty 101

The explicate order may be like ripples on the surface 102

Electrons and protons constructs of quarks 103

The cycle of plasma in the universe is a gigantic cycle 104

Does all this sound too exotic? I would say, it doesn't. 105

Actually, the plasma cycle is not exotic at all 106

The greatest existential challenge 107

Resonance effects of the intergalactic plasma streams 108

At the coldest level of the last 440 million years 109

We are now 2 million years into the ice age cycles epoch 110

Plasma streams have a built-in resonance 111

Solar cycles are essentially resonance features 112

The collapse of the primer fields happens at a lower density 113

High-powered mode for roughly 10% of the Ice Age Cycle 114

The interglacial period is asymmetric in nature 115

The interglacial is accelerating towards its end 116

Alarm bells should be ringing. 117

The first major effect that happened 118

We saw the shorter cycles diminishing likewise 119

Intervals getting shorter at a rate of geometric progression 120

Many of the dynamic features are presently fast loosing ground 121

Fringe-effect consequences on the wide field of climate effects 122

The greenhouse effect of the atmosphere getting weaker 123

Ulysses saw a 20% increase of cosmic-ray flux per decade 124

We now see the same fast rate continuing 125

Even the heart beat of the solar system is slowing down 126

Real-time measurements become important 127

When the primer fields collapse 128

The real, deep Ice Age Challenge stands before us all 129

The needed self-healing of humanity is possible 131

More Illustrated Science Books by Rolf A. F. Witzsche 132

False consensus: The Sun is a sphere of hydrogen gas

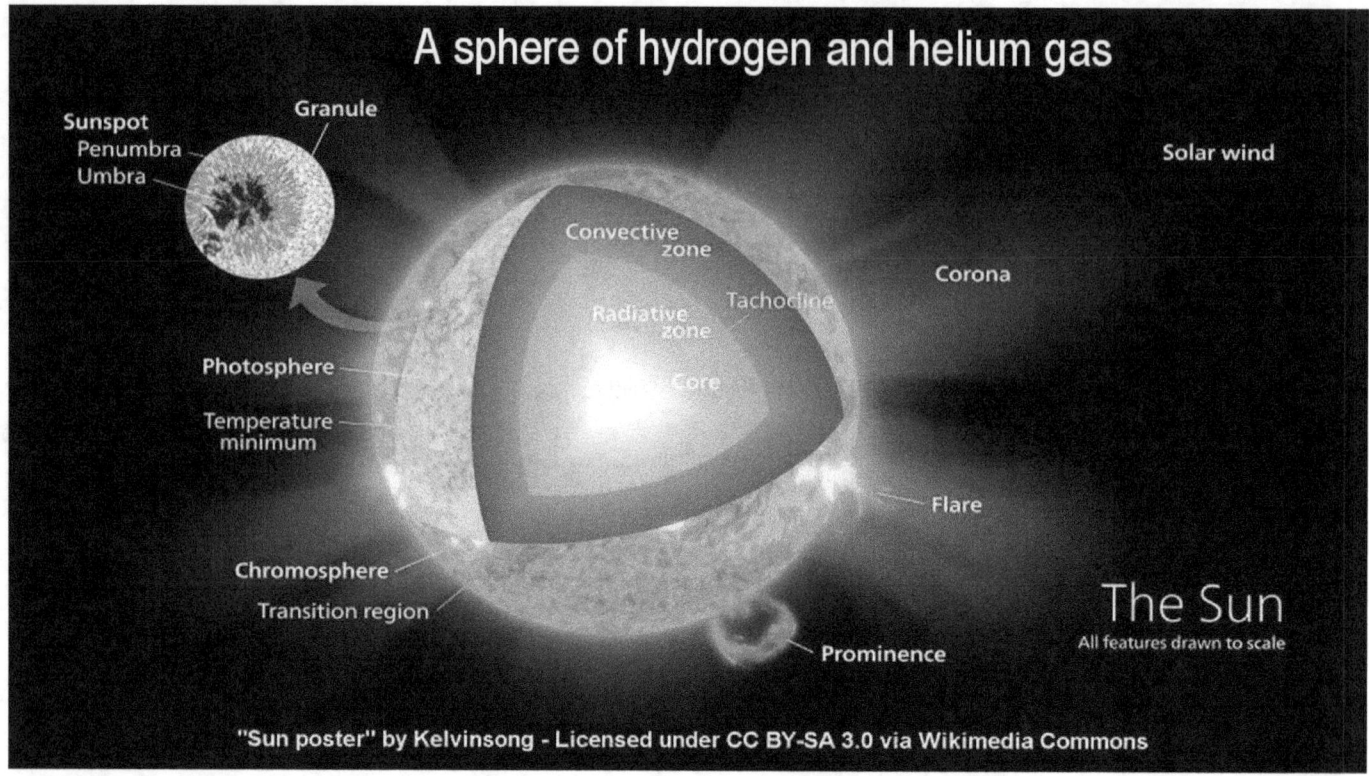

The mainstream science consensus is that the Sun is a sphere of hydrogen gas that is internally heated. Gravitational compression is deemed to have raised the temperature of its core to 15 million degrees, and to an unimaginable great mass density at its core, in the order of a 150 times the density of water. The temperature and pressure is deemed to be so great, that hydrogen atoms at its core are forced to combine into heavier atoms, primarily into helium atoms, and it is further deemed that the process liberates energy, which is deemed to be the energy that the Sun radiates. It deemed that the energy generated reaches the surface over a span on 10,000 years, to 170,000 years, or up to 30 million years as it was once believed.

Whatever the case may be, the widely accepted theory of the nature of the Sun is full of holes and paradoxes, and is devoid evidence that exclusively supports the theory. Let's look at the paradoxes.

The Sunspot Paradox

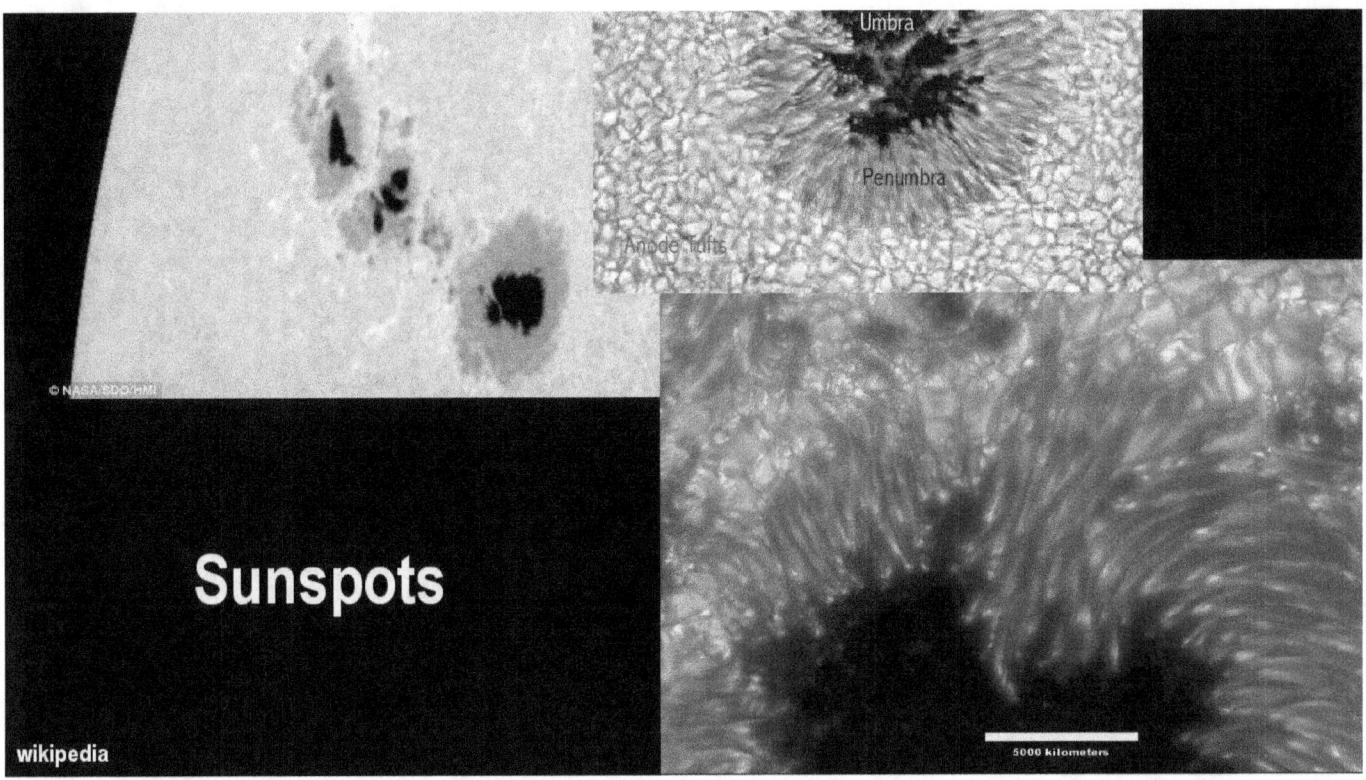

** The Sunspot Paradox

Paradox #1 is that the observed sunspots, which are holes in the Sun's surface, are dark spots, not bright spots as they should be if the Sun was internally heated.

The dark umbra of the sunspots is theorized 'dark energy'

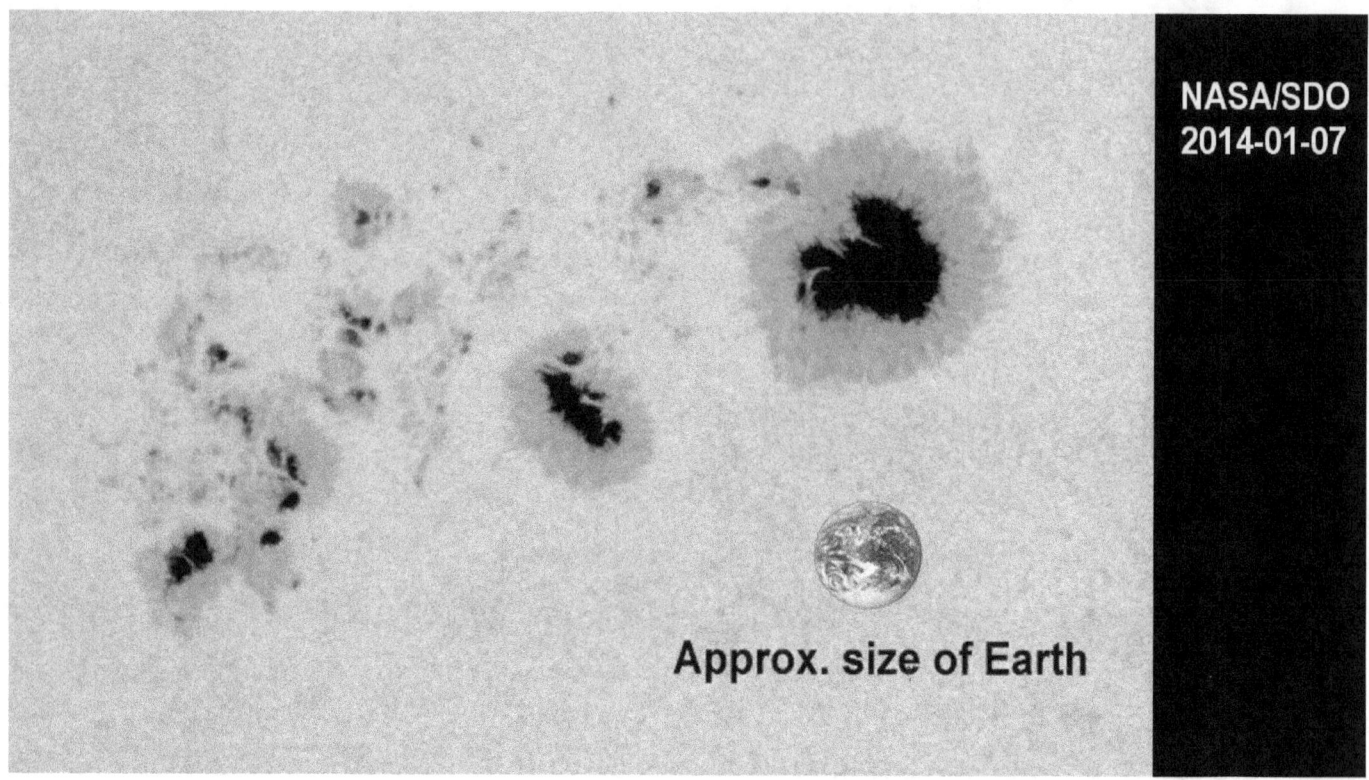

The dark umbra of the sunspots is theorized to result from 'dark energy' streaming through the photosphere, blocking the light from within, which is a paradox in itself.

The Solar-Wind Paradox

- an example of the amazing solar eclipse photography of Milloslav Druckmueller

** The Solar-Wind Paradox

Paradox #2 is that the Sun emits streams of free plasma particles as solar wind. The solar wind of plasma particles shouldn't be happening when the nuclear fusion process deep inside the Sun is said to combine atoms into larger ones, rather than tearing them apart to produce plasma.

The Solar-Wind-Acceleration Paradox

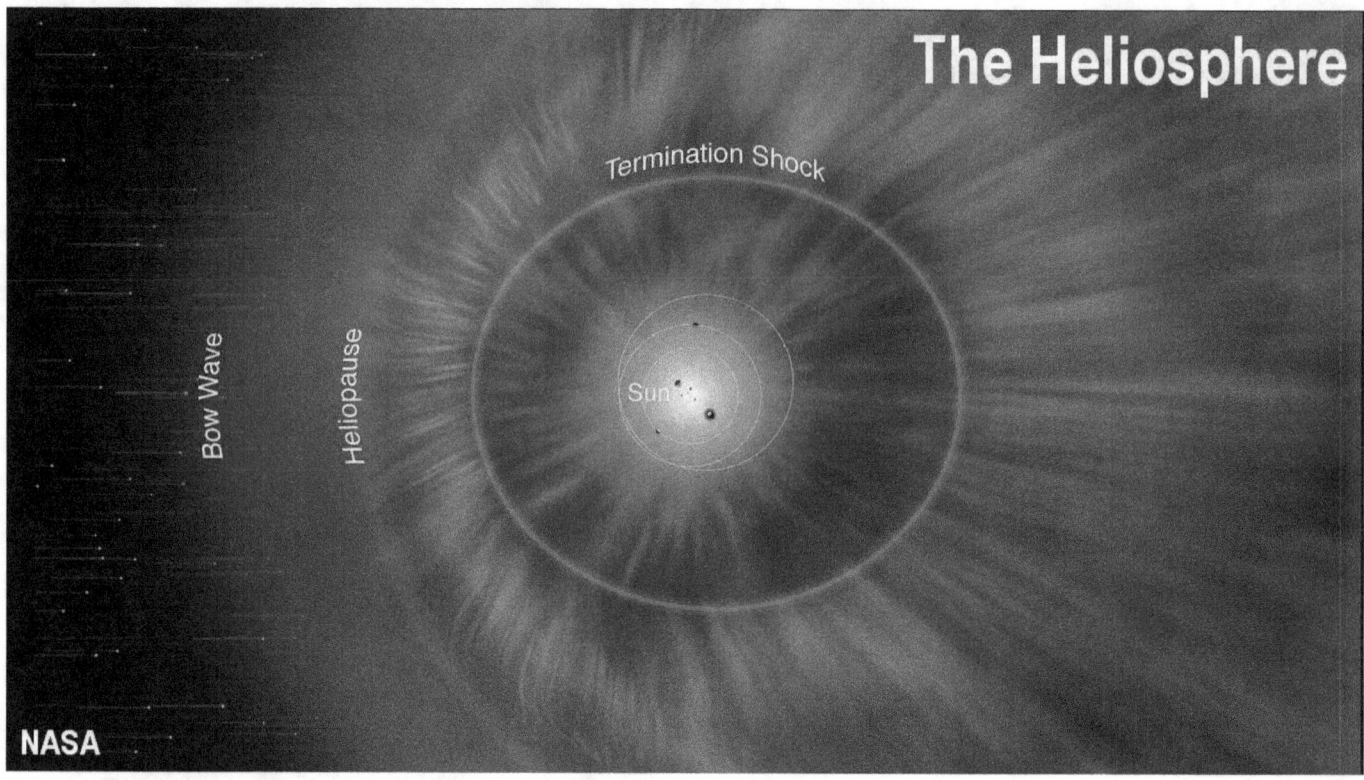

** The Solar-Wind-Acceleration Paradox

Paradox #3 is that the solar wind accelerates as it flows away from the Sun, against the force of the Sun's gravity, up to a distance 100-times greater than the distance from the Sun to the Earth, where it forms the heliosphere.

The Super-Hot-Corona Paradox

** The Super-Hot-Corona Paradox

Paradox #4 is that the Sun's surrounding corona is enormously hotter than the Sun itself. This shouldn't be possible when the heat comes from within. The outside shouldn't be hotter than the inside.

The Low-Mass-Density Paradox

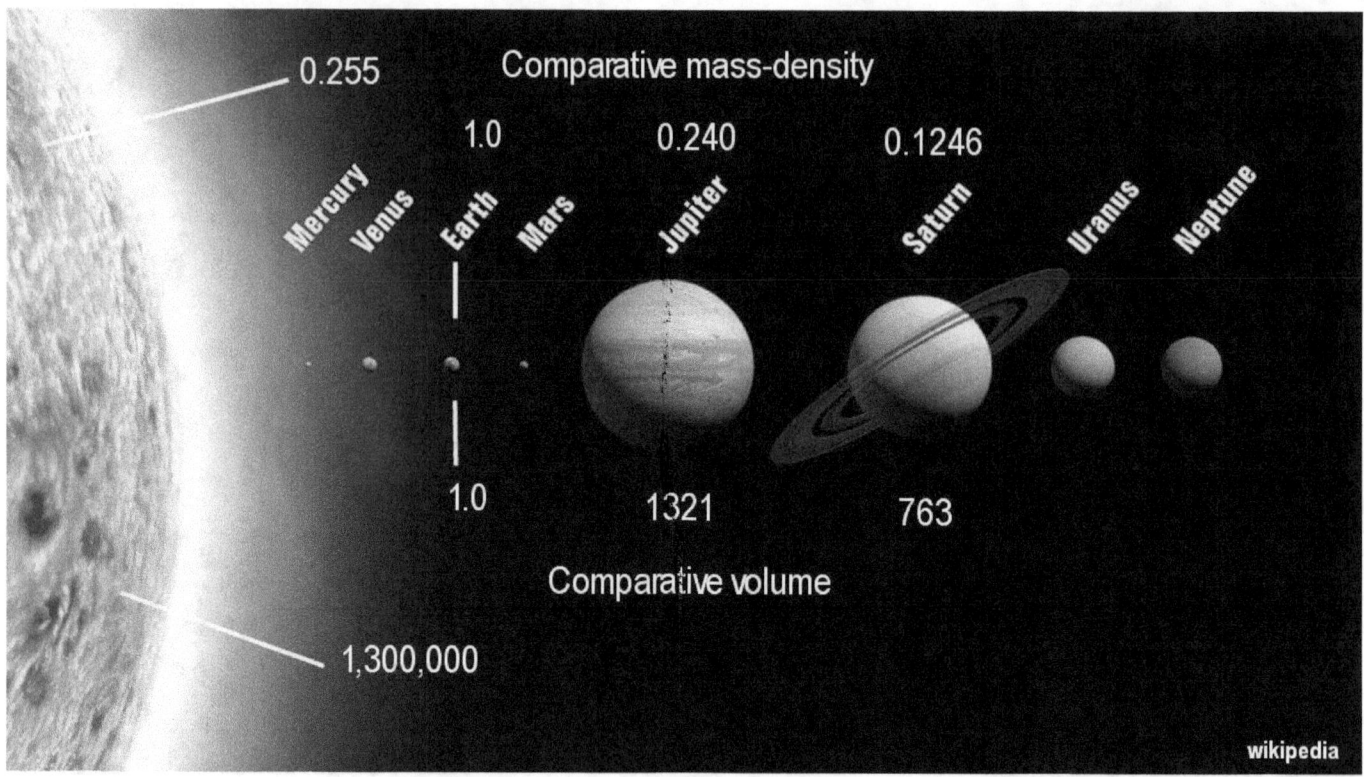

** The Low-Mass-Density Paradox

Paradox #5 is that the Sun is a thousand times too light, for a gas sphere of its size, if one compares the Sun with known gas spheres. With Jupiter being a gas planet of twice the volume of Saturn, Jupiter weighs in with twice the overall mass density than Saturn as the natural consequence of greater gas compression by Jupiter's greater gravity. But if one extends the comparison to the Sun, which has a thousand times larger volume than Jupiter, the principle of gas compression isn't reflected. The Sun weighs in with roughly the same overall mass density than Jupiter. That's a paradox. The Sun is a thousand times too light for a gas sphere of its size, especially with a compressed core that is 150 times denser than water. The measured facts defy the gas-sun theory.

The Giant-Stars Paradox

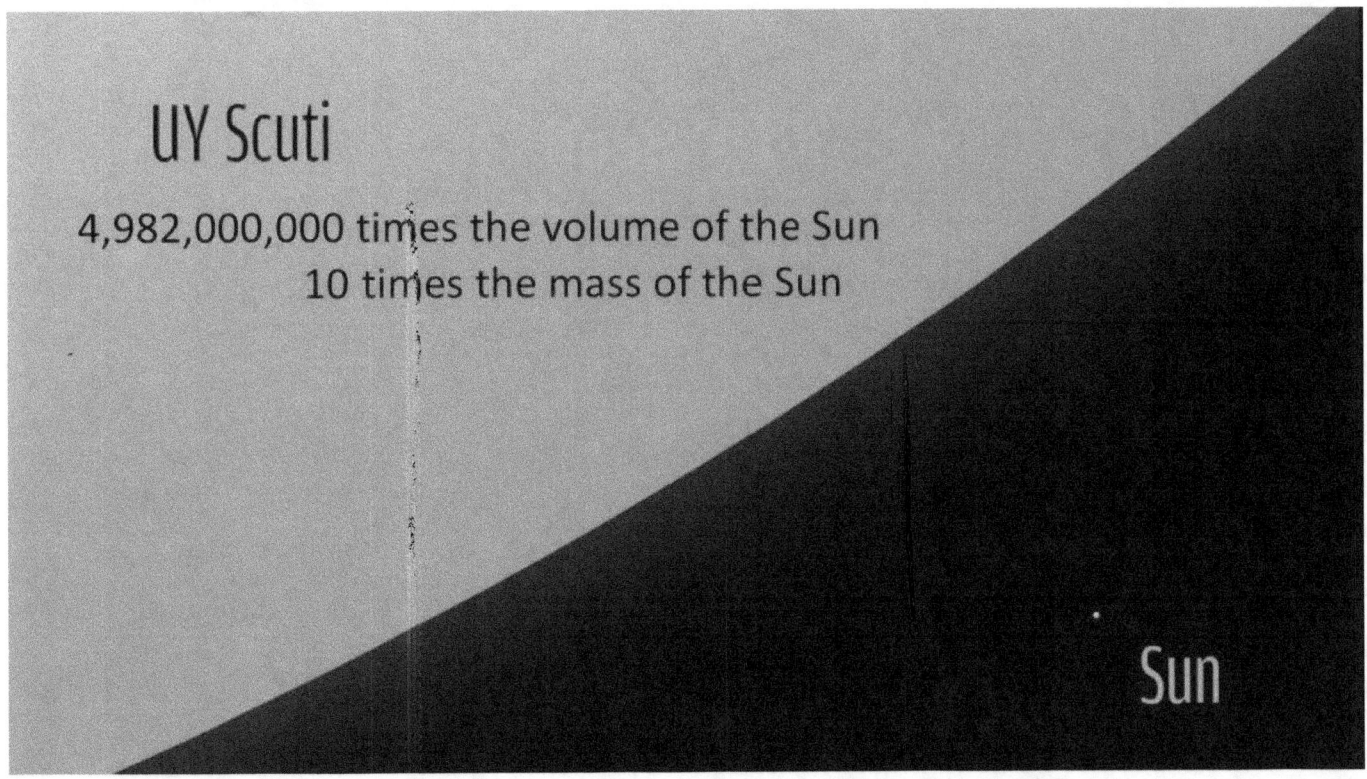

** The Giant-Stars Paradox

Paradox #6 is that giant stars shouldn't exist, like the star UY Scuti. This star has a five billion-times greater volume than the Sun, but only contains 10 times the mass of the Sun. This shouldn't be possible under the gas compression model.

This star exists, and it outshines our Sun 340,000-fold

But this star exists, and it outshines our Sun 340,000-fold. This too, shouldn't be possible under the gas compression model for the Sun.

The Solar Cosmic-Rays Paradox

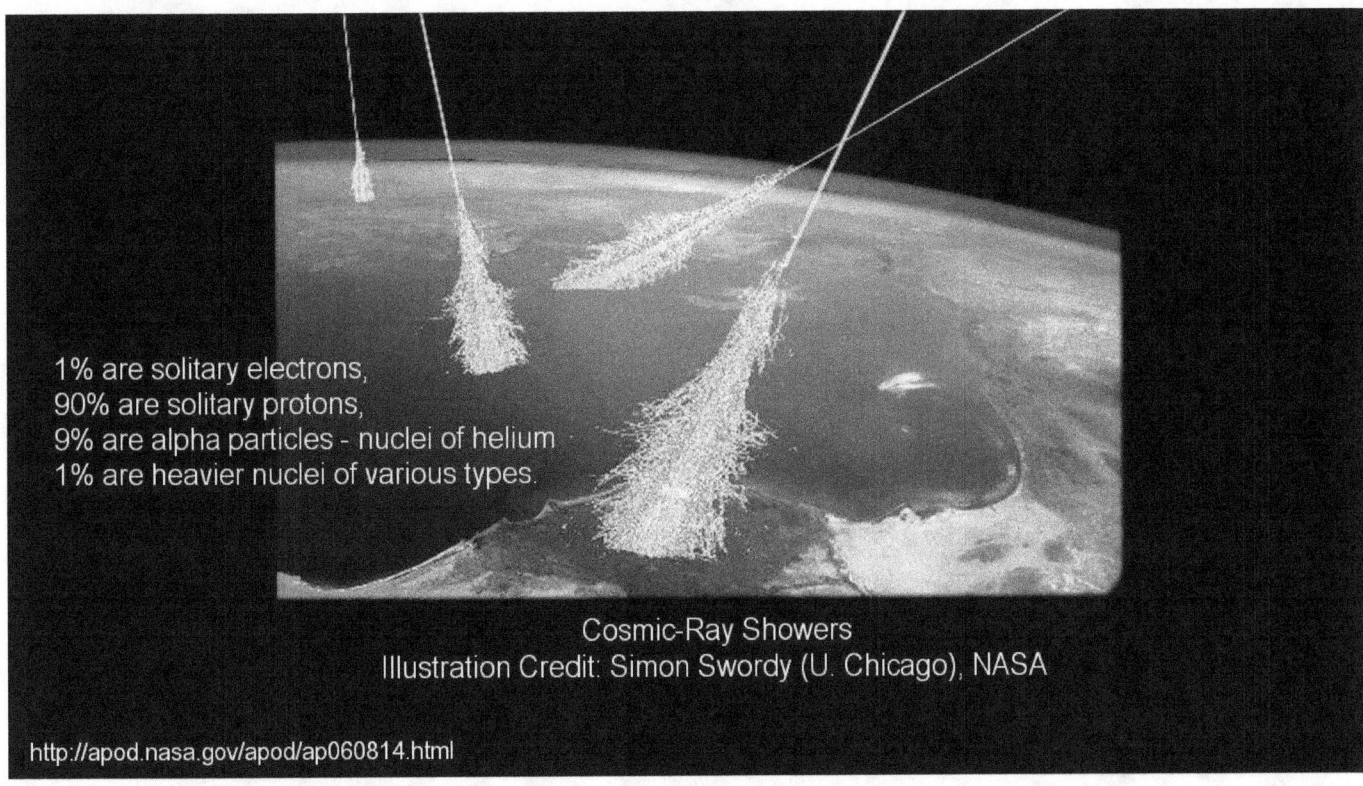

Cosmic-Ray Showers
Illustration Credit: Simon Swordy (U. Chicago), NASA

** The Solar Cosmic-Rays Paradox

Paradox #7 is that the Sun also emits cosmic-ray flux. Cosmic rays are single events of highly energized protons and electrons.

It is deemed impossible for these to originate from a gas sun, so that all cosmic-ray events are deemed to be galactic in origin. It is reasoned that if the cosmic-ray particles get intercepted in the Earth's thin atmosphere, that's just a few kilometers deep, none would make it past the Sun's half-million kilometers deep outer layer of hydrogen gas.

Cosmic-ray fluctuations with the Sun's activity cycles

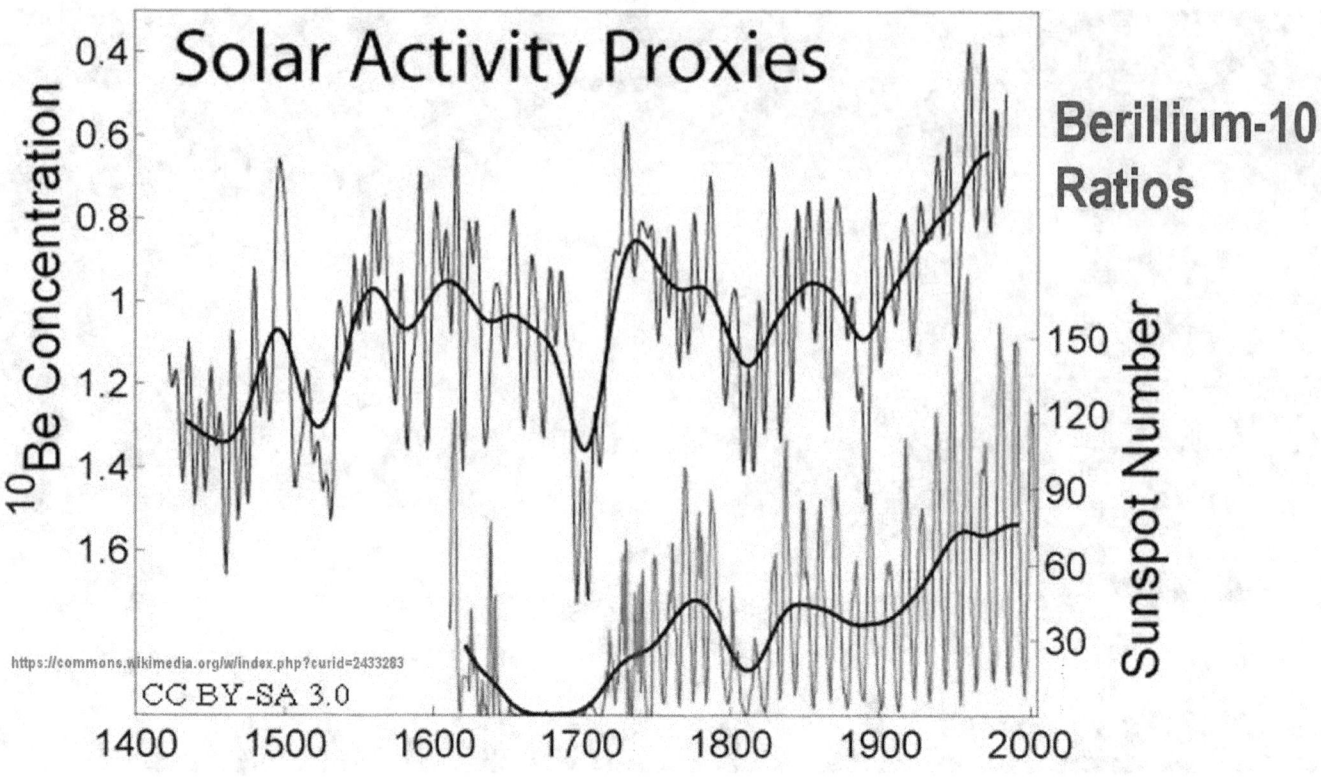

The paradox here is that cosmic-ray fluctuations have been measured in proxy to consistently fluctuate in perfect synchronism with the Sun's activity cycles, as it is shown here in Berillium-10 ratios.

Synchronism measured by the Moscow Neutron Monitor

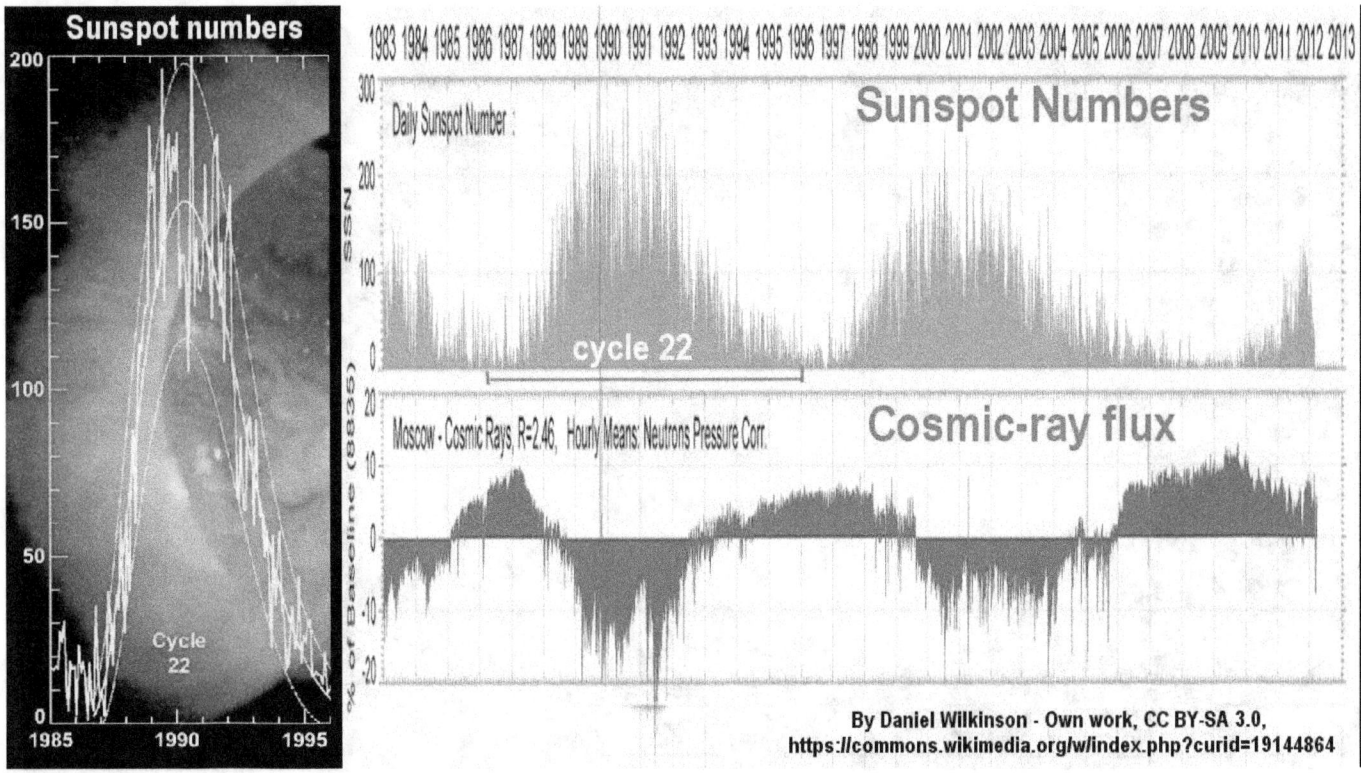

The same lock step synchronism of cosmic-ray flux with the solar cycles is also being measured in real time by the Moscow Neutron Monitor.

The Solar-Cycles Paradox

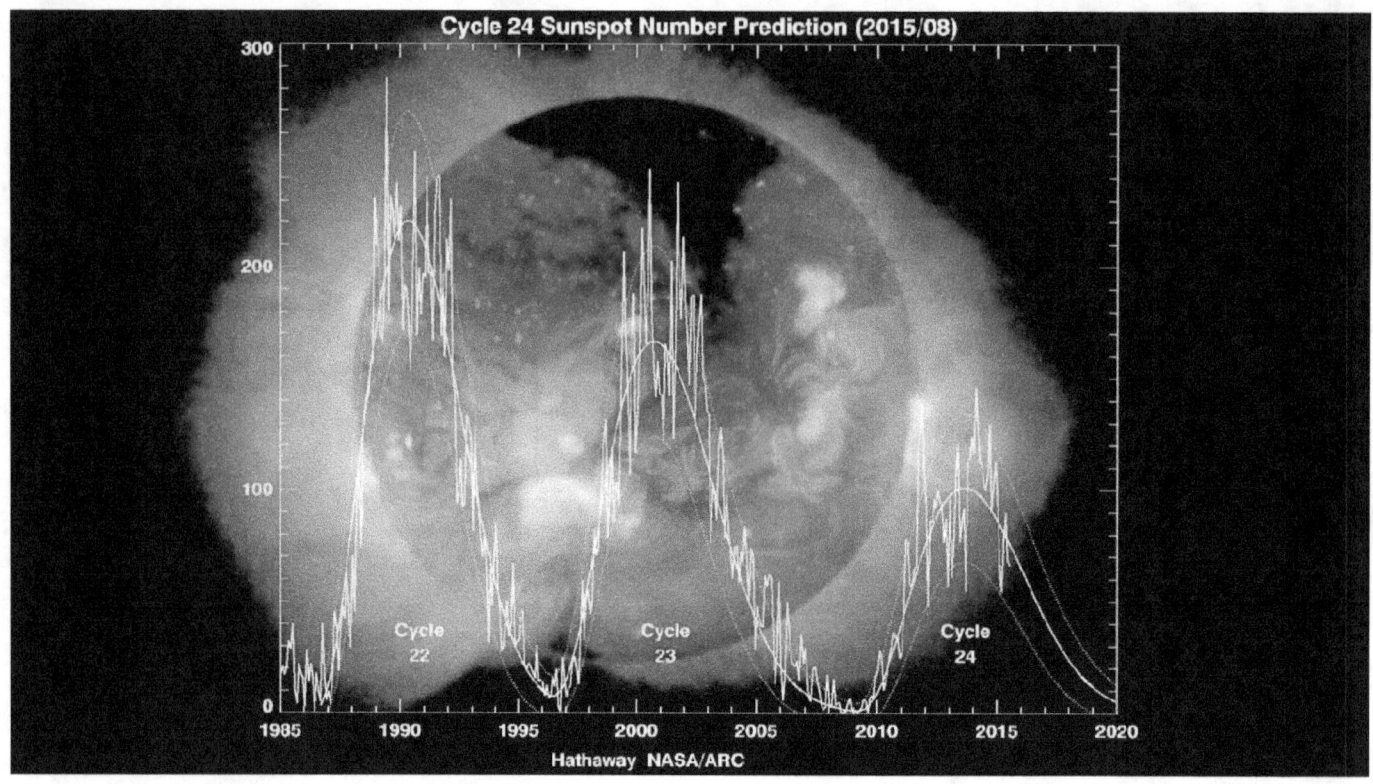

** The Solar-Cycles Paradox

Paradox #8 is that the Sun's activity is pulsing in 11-year solar cycles, and that the pulsing is changing in intensity. This fast pulsing and diminishing shouldn't be possible for a gas sun in which the solar energy takes 10,000 years, up to 170,000 years or more, to reach the surface.

The Sunlight Paradox

** The Sunlight Paradox

Paradox #9 is the sunlight that the Sun emits. With a few exceptions, all visible light is emitted from energized atomic elements. The light from atoms is emitted in different bands within the visible spectrum, according to the physical characteristic of the emitting atomic structures.

The Sun emits a richly homogenous spectrum

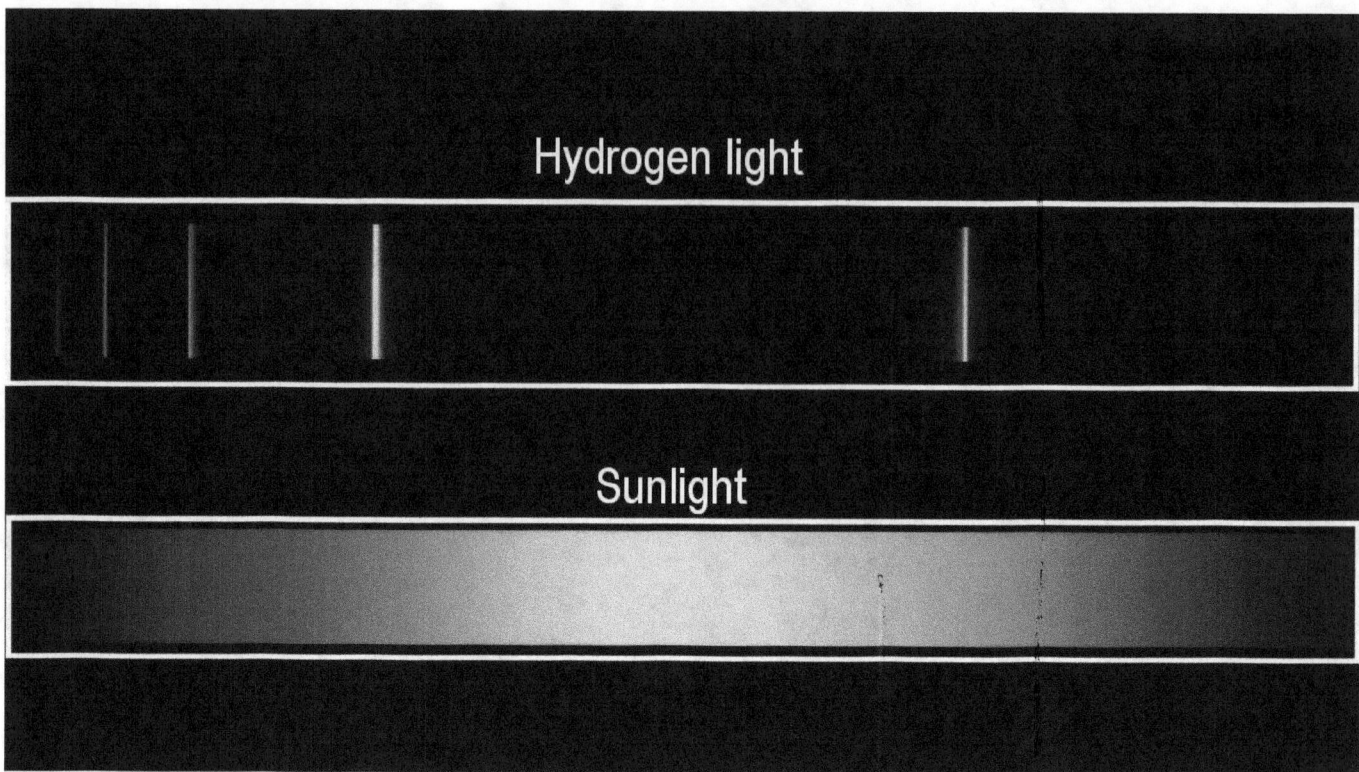

Since the Sun is deemed to be a hydrogen gas star, its emission spectrum should match the emission spectrum of hydrogen gas. But this is not the case. It is well known that the Sun emits a richly homogenous spectrum, and that this kind of spectrum can't possibly originate from a hydrogen gas star that our Sun is deemed to be. That's a paradox.

The Atomic Elements Paradox

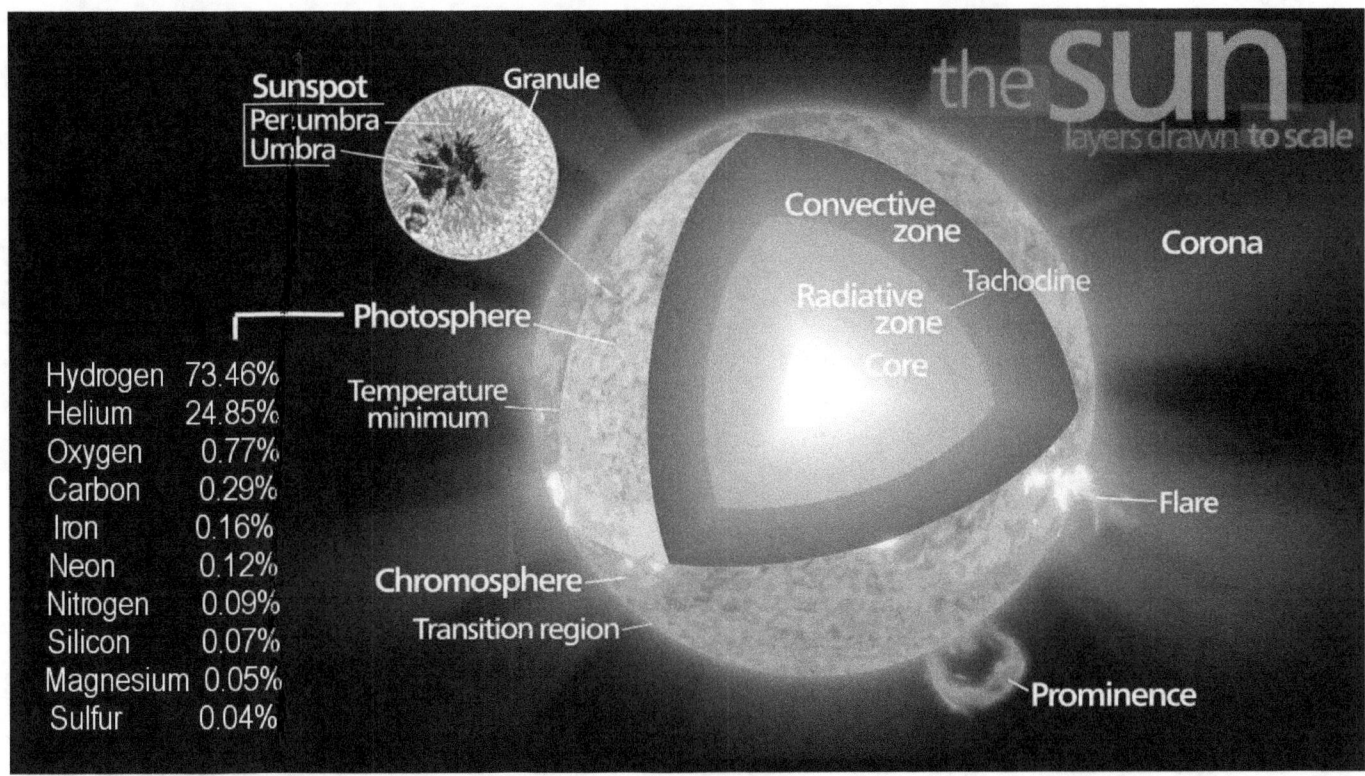

** The Atomic Elements Paradox

Paradox #10 is, that the Sun is known to be surrounded with a wide rage of atomic elements, which have been detected by their spectrum in the photosphere. Paradoxically, they shouldn't be found on the surface of the hydrogen Sun. These heavy elements should have been purged from the surface by the solar wind, or been swallowed by the Sun's gravity, especially the heavy, iron.

The Differential Rotation Paradox

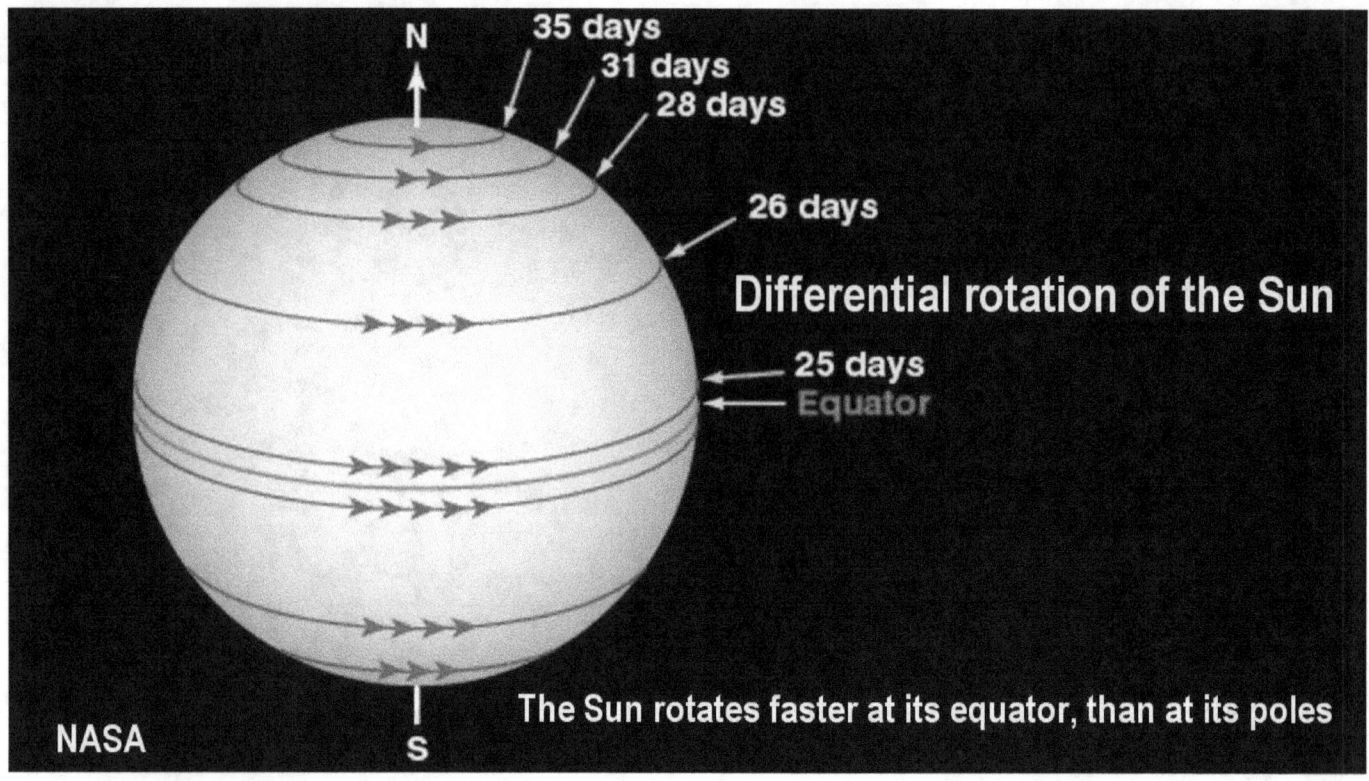

** The Differential Rotation Paradox

Paradox #11 is that the Sun rotates faster at its equator than at its poles. This isn't possible for a gas sphere. Still it is happening.

Consensus model for the Sun is fundamentally wrong

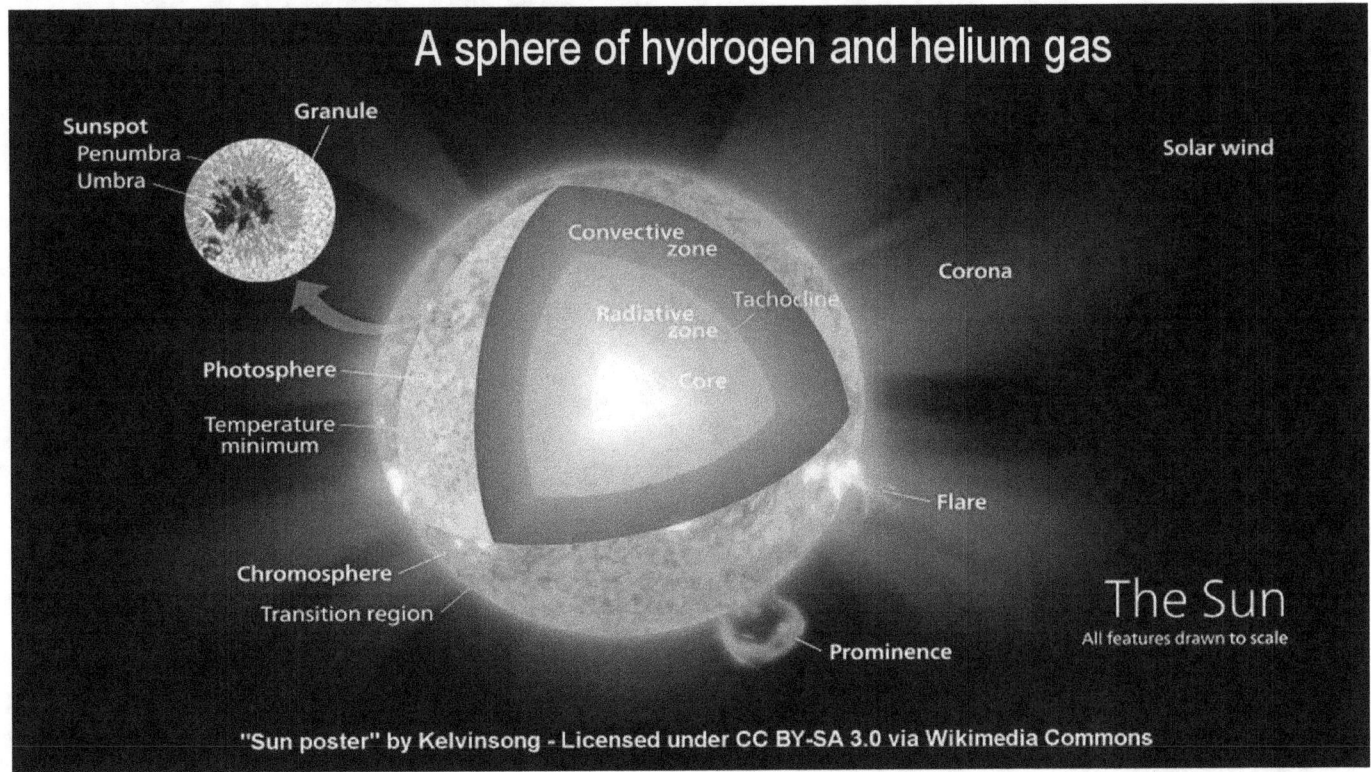

These many paradoxes suggest rather strongly that the mainstream consensus model for the Sun is fundamentally wrong.

This also means that a different model needs to be developed that is free of those paradoxes, and which in addition, is supported by measurable physical evidence.

NASA has helped us in this quest.

Differences in Mass Distribution

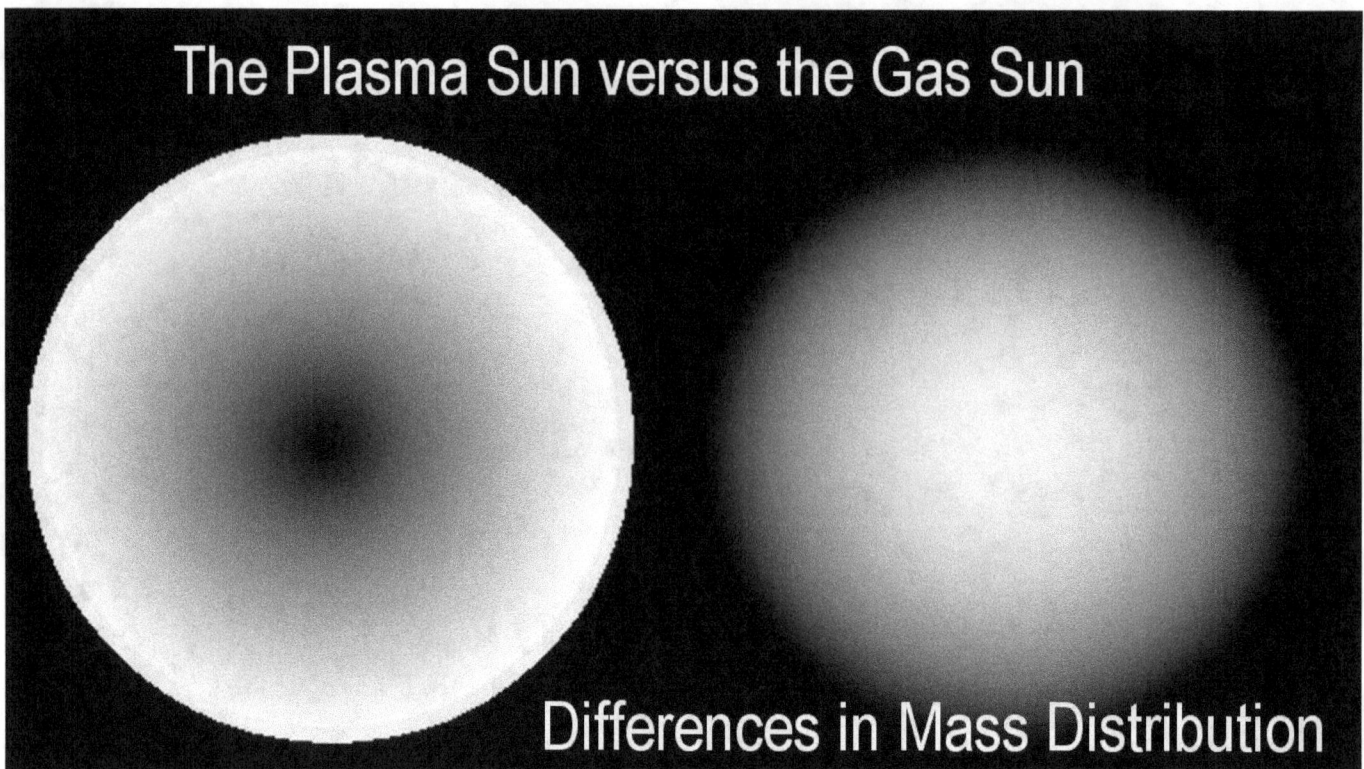

The Plasma Sun versus the Gas Sun.

Differences in Mass Distribution

NASA has opened the portal to us, that enables us to see the Sun as being radically different than it is deemed to be. NASA's discoveries enable us to see our Sun as being a Plasma Star, with qualities that make it a natural sun.

With this in mind, let's explore the concept of the Plasma Star concept. Let's do this first so that NASA's revolutionary contributions can be more easily recognized.

The Plasma Star: A leading-edge concept in astrophysics

The Plasma Star: A leading-edge concept in astrophysics.

A plasma star is a large sphere of plasma

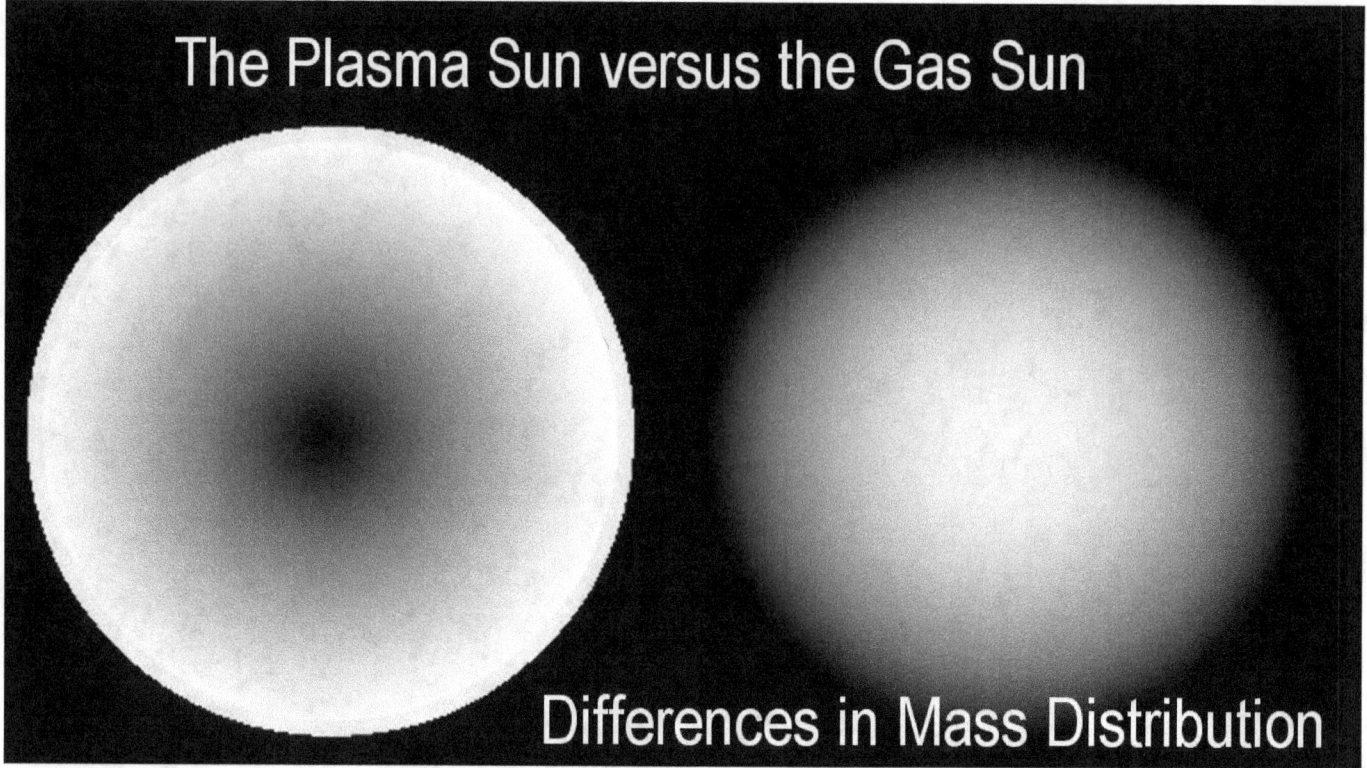

A plasma star is a large sphere of plasma, with differences in mass-distribution that give the plasma sphere a radically different characteristic than that of a gas sphere.

Plasma is the name for subatomic particles

**atoms are formed by the dynamic 'dance'
of electrons being attracted and forced to rebound**

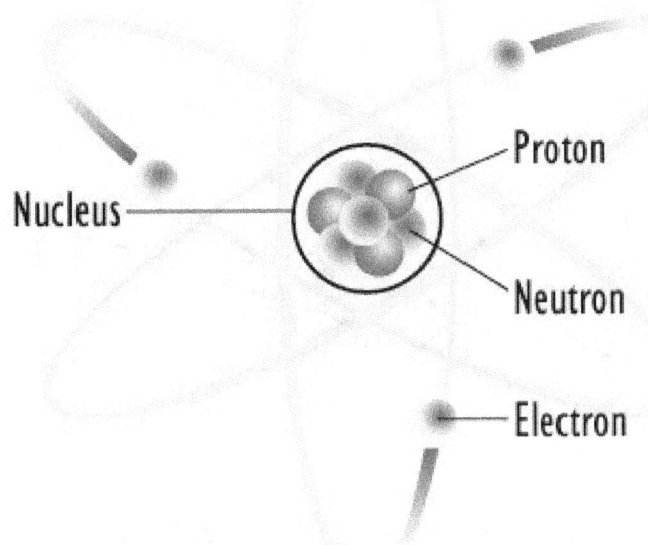

wikipedia (image)

Plasma is the name for subatomic particles when they exist in space in free form, instead of being bound into atoms.

When plasma particles are free

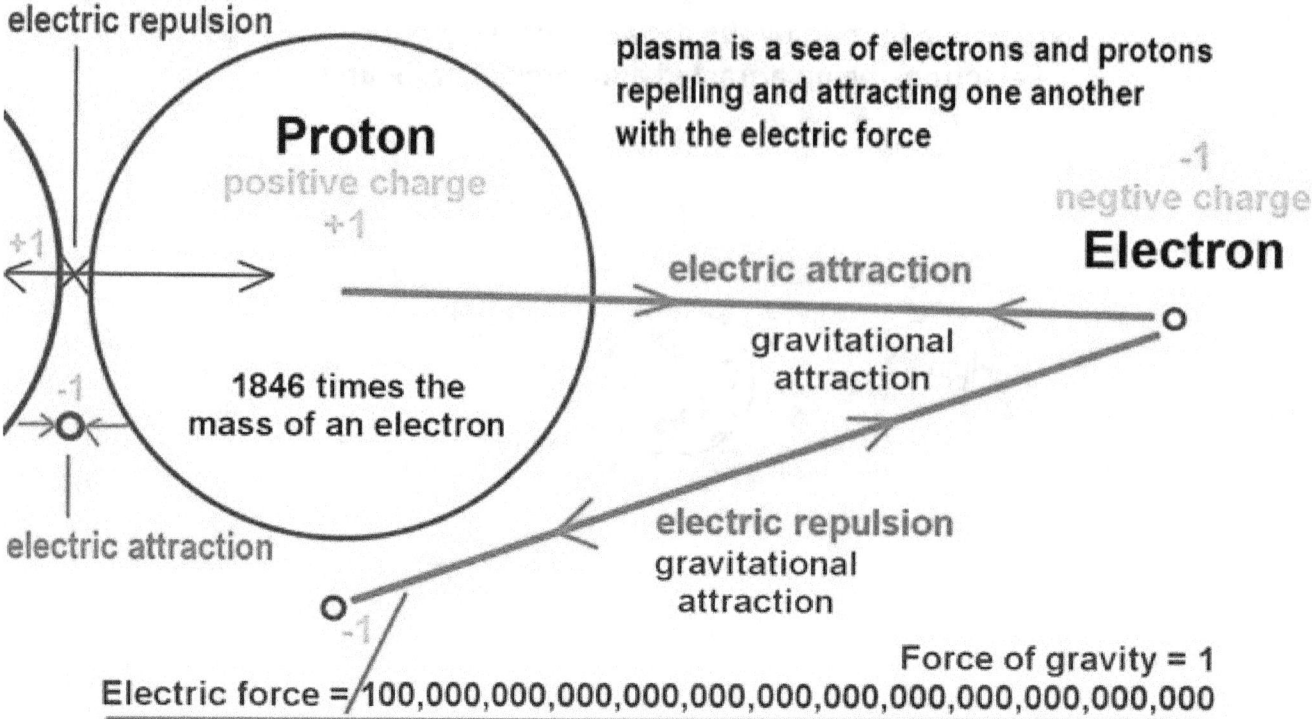

When plasma particles are free, the individual plasma particles radiate their electric potentials. This means they interact with each other with one of the strongest forces in the universe, which is the electric force that is 39 orders of magnitude stronger than the force of gravity.

There exist two types of plasma particles. One type is named the electron. It carries a negative electric potential. And the other type, which is 1800 times larger in size, is named the proton. It carries a positive electric potential.

Particles of like potential repel each other, and those of unlike potentials attract each other.

In plasma, the tiny electron is intensely drawn to the big proton

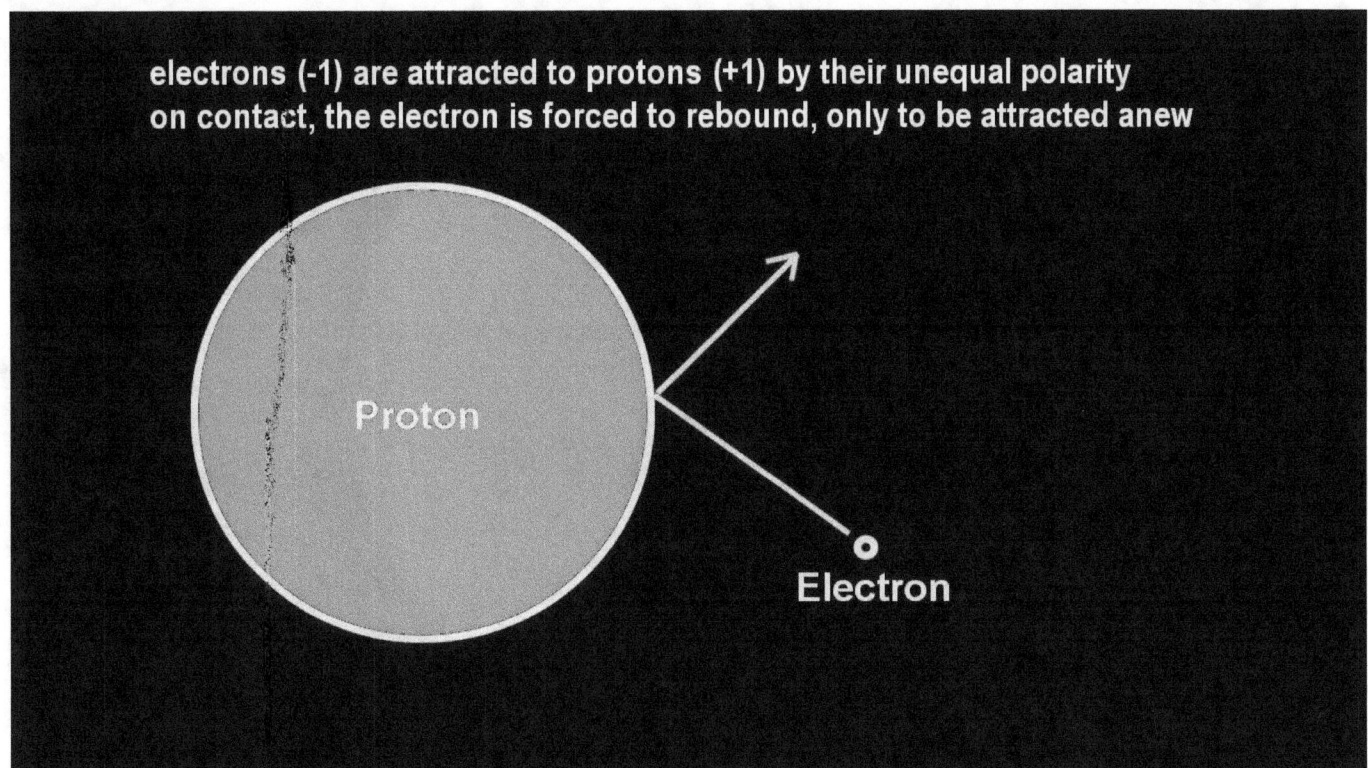

In plasma, the tiny electron is intensely drawn to the big proton. But before the electron can latch itself onto the proton, the electron is repelled at a close distance by a unique nuclear force. The repelled electron, of course, becomes attracted anew by the electric force. In this manner, the electrons in plasma are drawn into an endless dance around the protons.

A large plasma sphere has its least mass density at its core

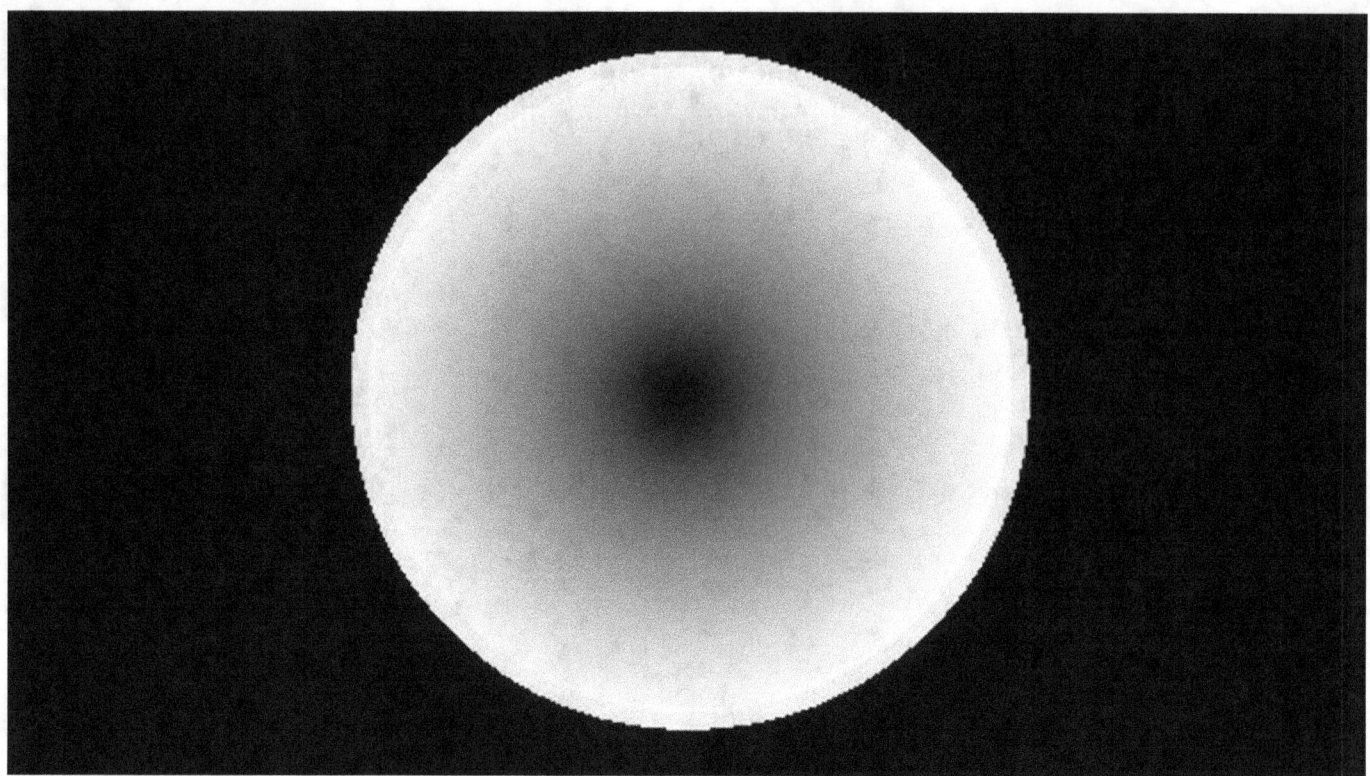

But within a large sphere of plasma, the force of gravity also plays a role. By gravitational pressure, the dance of the electrons tend to migrate the electrons away from the center of gravity, towards the surface. This means that protons at the core of a large plasma sphere are being less swarmed about. They become thereby less eclectically isolated, and more able to repel each other with the electric force.

The end-result is that a large plasma sphere has its least mass density at its core, and its greatest mass density at its surface, and with it also its greatest electron density.

With its electron-rich surface, a large plasma sphere is a natural sun

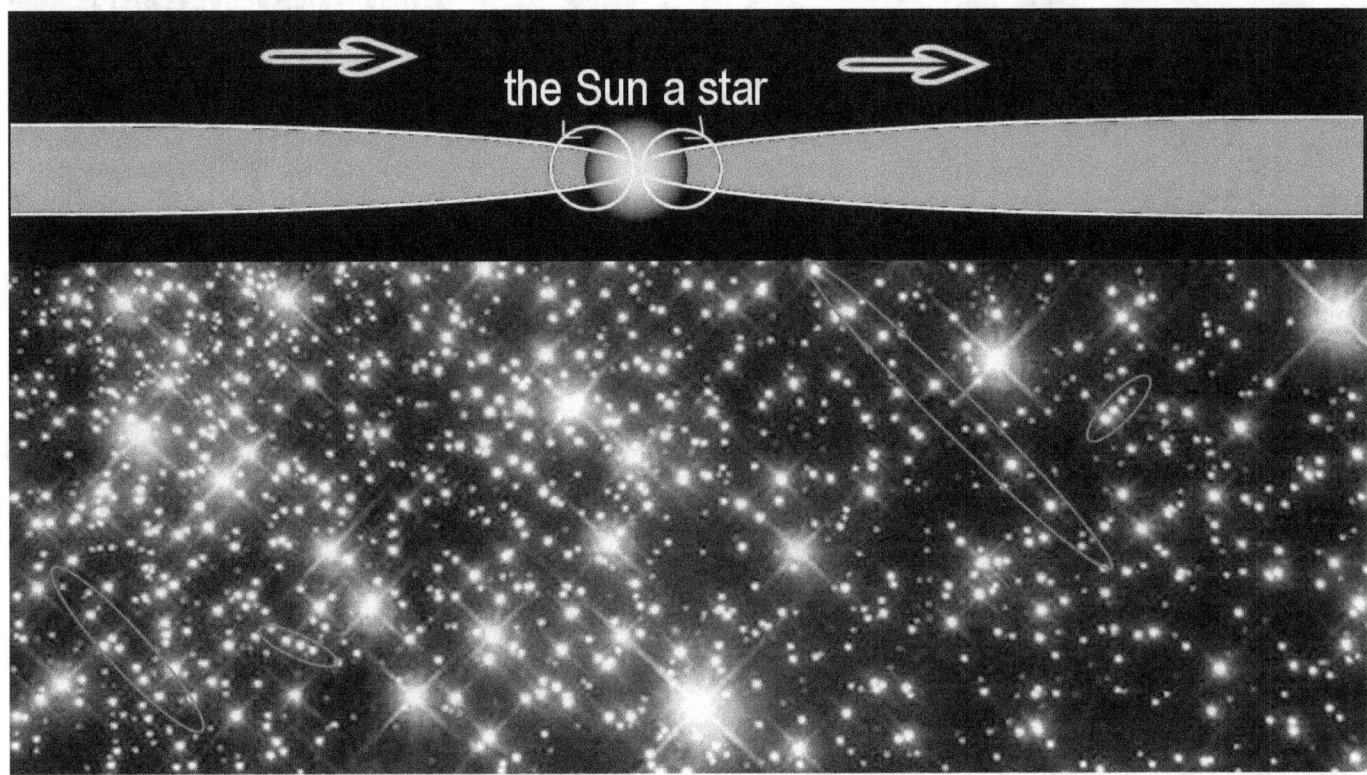

With its electron-rich surface, a large plasma sphere is a natural sun. It is a sun, because its highly concentrated negative electric potential radiates far into space and attracts interstellar plasma to it, that it interacts with in such an energetic manner that the inflowing plasma particles become bound to each other by organizing principles with which all known natural atomic elements are synthesized.

A plasma sphere is a plasma star

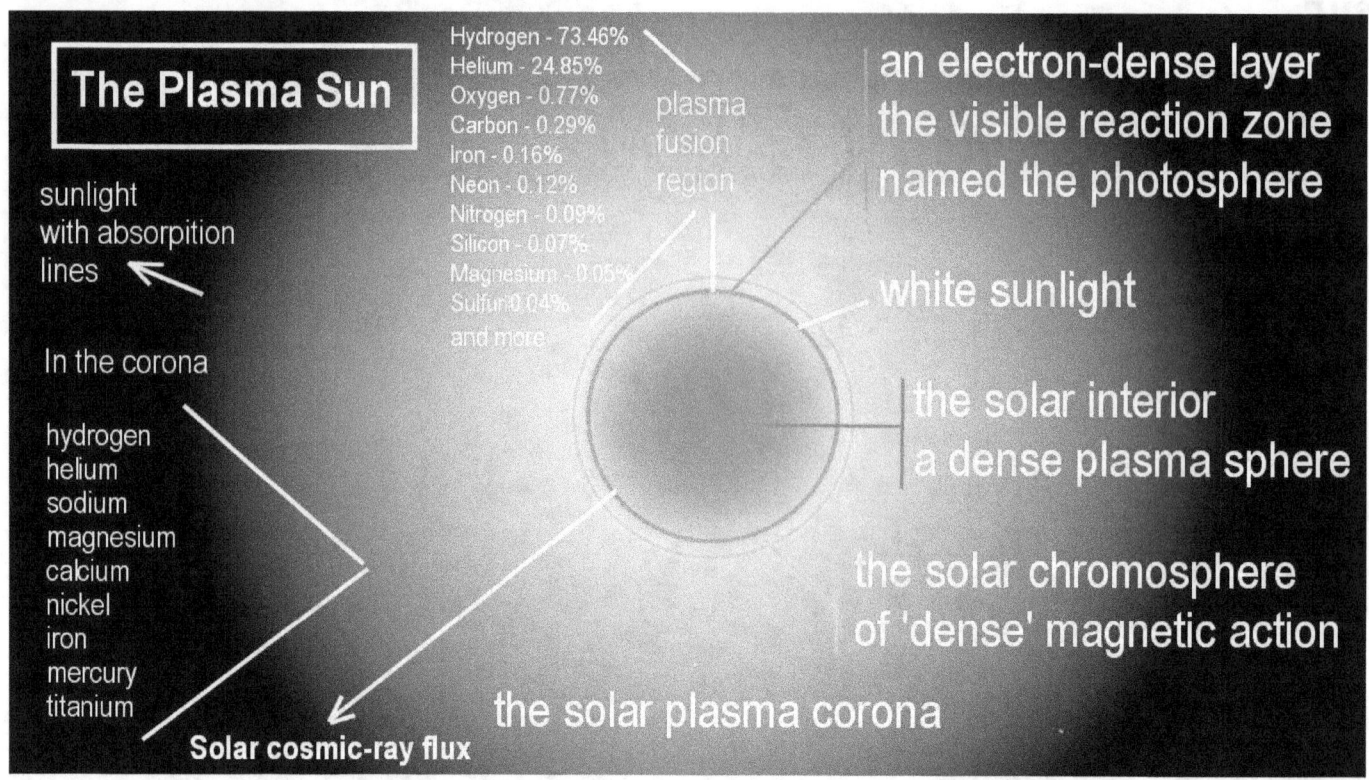

By this process, a plasma sphere is a plasma star.

A plasma star is essentially a largely empty sphere of plasma

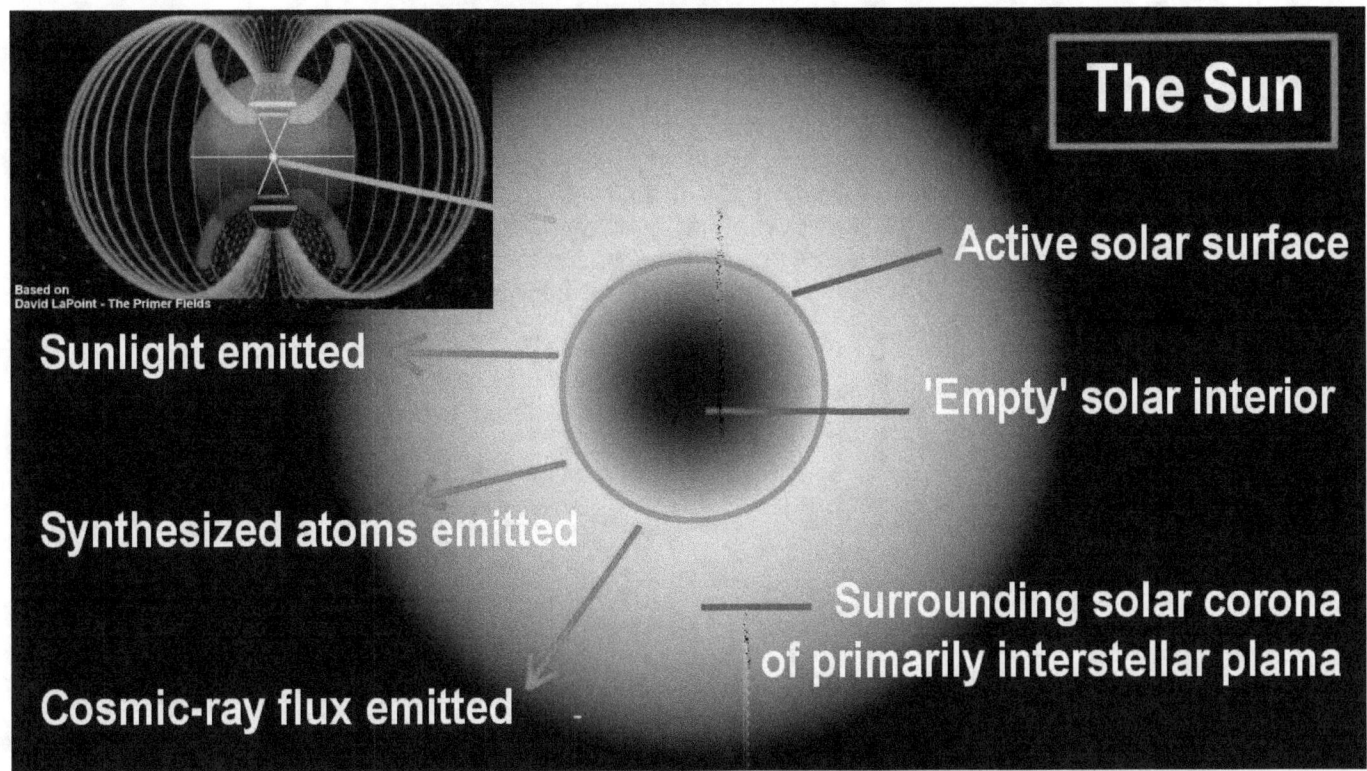

A plasma star is essentially a largely empty sphere of plasma, with an intensely active surface region where all the energetic actions take place, which in turn is surrounded by a sphere of interstellar plasma that it attracts, that becomes concentrated around it by a complex electromagnetic process.

This means that nothing actually happens inside a plasma sun, which our Sun is, because it fits all the parameters.

Sun's energy is generated on its surface

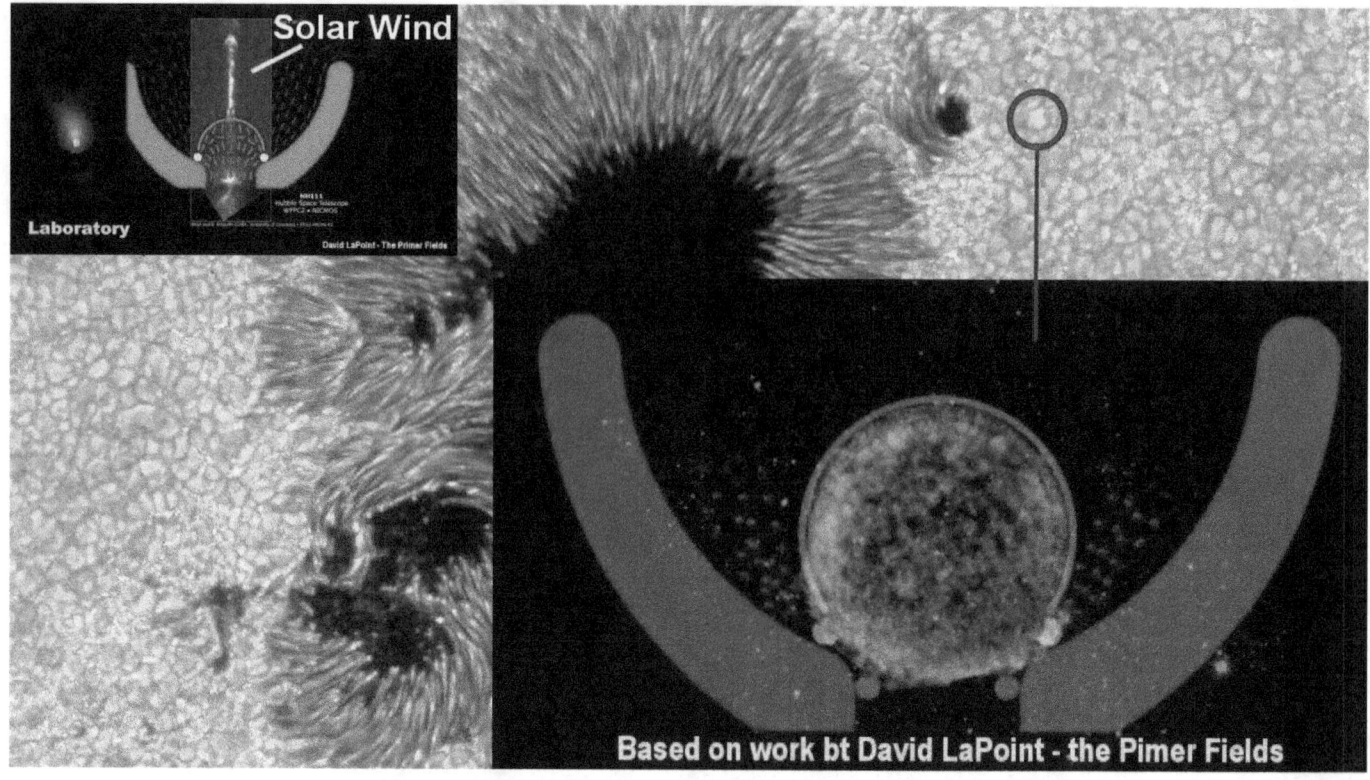

Our plasma-Sun's energy is generated on its surface in a sea of reactions cells where plasma is intensely fused into atomic elements of all types that are known to exist naturally.

A wide emission spectrum

Of course, the highly energized synthesized atomic elements emit large volumes of light and with a wide emission spectrum, because a large variety of atoms are involved in the light-emitting process.

The Plasma Sun concept eliminates ALL paradoxes

The Plasma Sun concept eliminates ALL paradoxes of the Gas Sun theory of mainstream consensus.

The Plasma Sun solves Paradox #1

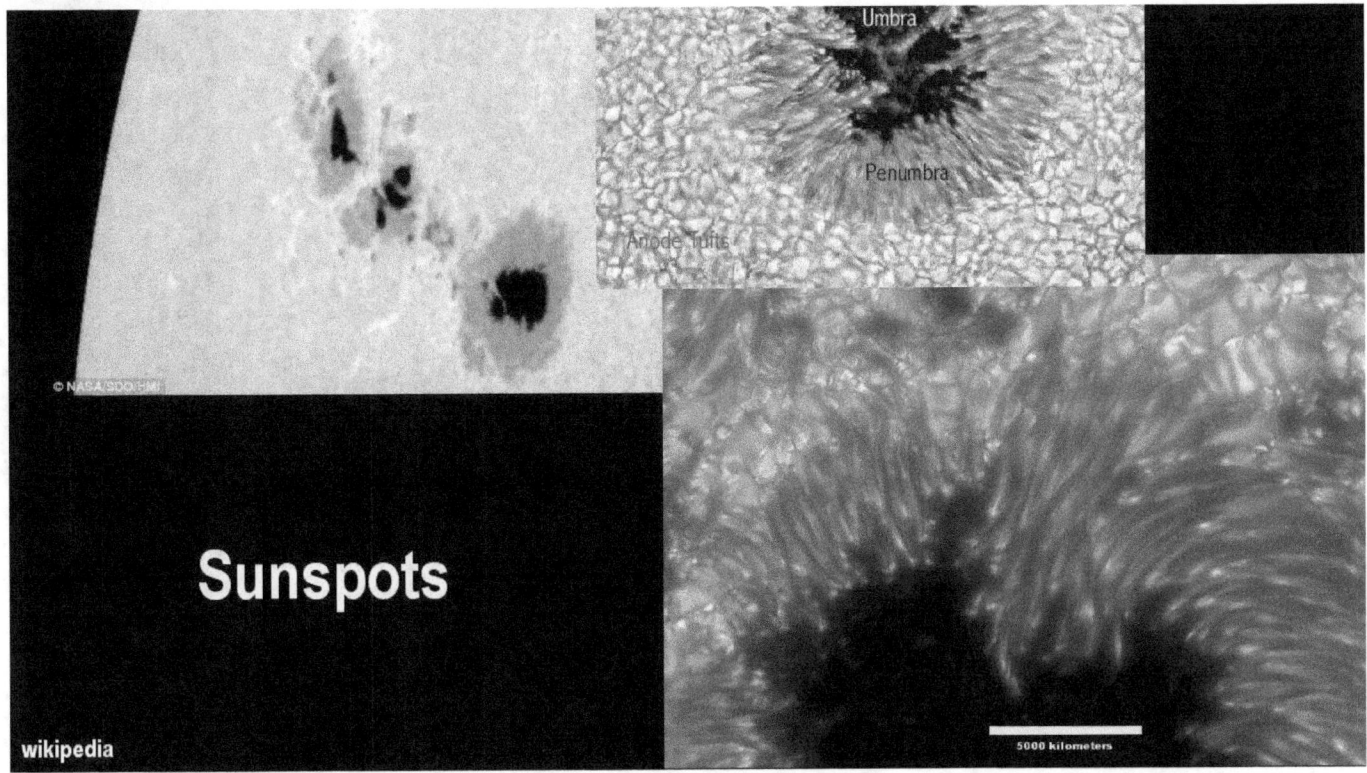

The Plasma Sun solves Paradox #1 by simply acknowledging that the sunspot umbra is dark, because nothing happens inside the plasma sun.

A plasma sun doesn't actually create energy

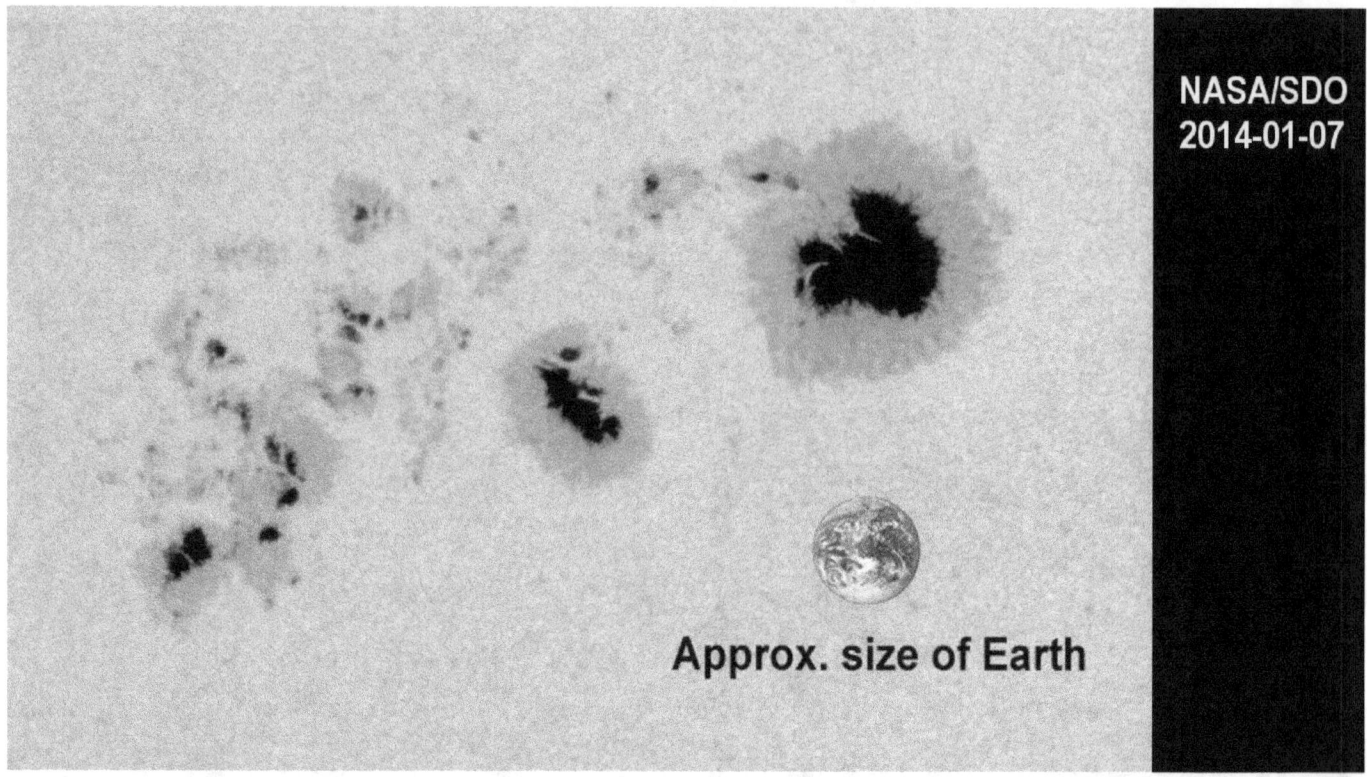

A plasma sun doesn't actually create energy. It merely attracts plasma and facilitates the plasma fusion that creates atomic elements in an energetic process on its surface that radiate some energy in the form of light.

Sunspots form when overload conditions cause the fusion cells to rupture, revealing thereby that the energetic solar action process is but skin deep.

Paradox #2, the solar-wind paradox

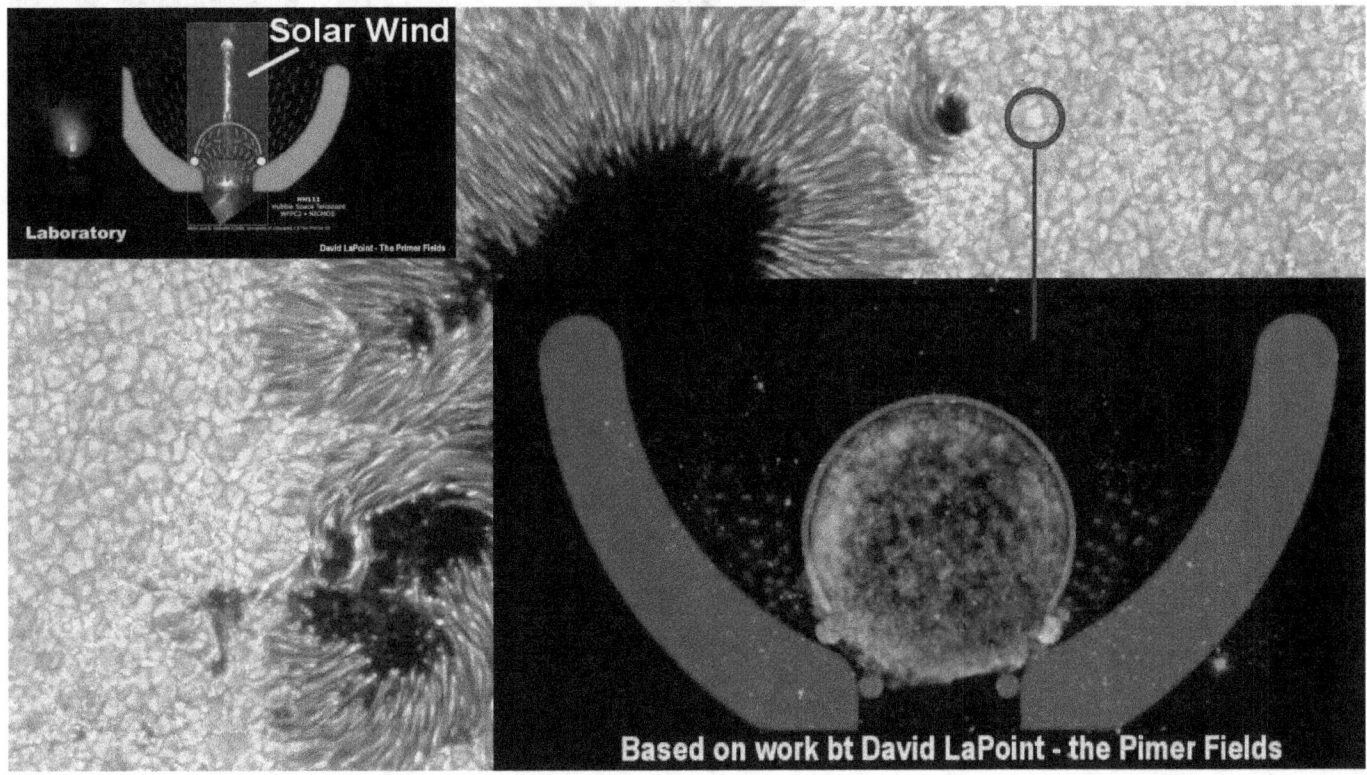

Paradox #2, the solar-wind paradox, is resolved by the build-in dynamics of the fusion cells.

Solar wind is only possible by a plasma-fusion process

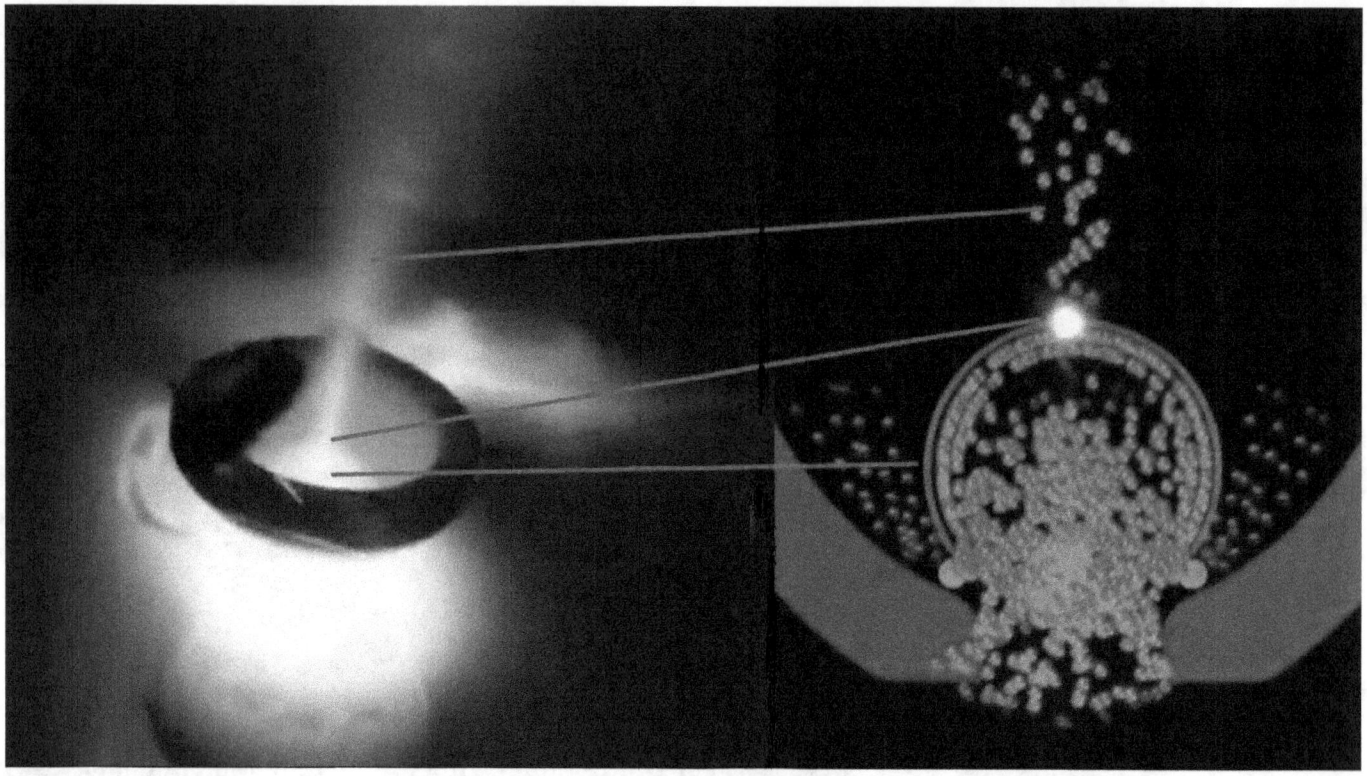

The in-flowing plasma becomes intensely concentrated by magnetic fields that are generated by the movement of plasma. The movement of electric particles generates magnetic fields. When the confined pressure exceeds the strength of the confining magnetic fields, a portion of the concentrated plasma escaped at the weakest point at the top of the confining field.

This dynamic process has been experimentally verified by researcher David LaPoint.

The escaping plasma from the top of fusion cells becomes collectively the solar wind. Solar wind is only possible by a plasma-fusion process, and this occurs only on the surface of a plasma Sun.

Paradox #3, the solar-wind acceleration paradox

Plasma escaping a magnetic confinement structure.

Paradox #3, the solar-wind acceleration paradox, is not a paradox for the plasma sun, because the solar wind particles are emitted in highly concentrated form, from the fusion cells. When intensely concentrated plasma escapes the magnetic confinement structure, it becomes free to expand again, which it does explosively by the electric force. The flowing plasma, of course, forms magnetic fields around it that keep the flowing stream pinched together, within which that plasma continues to expand. This, too, has been verified experimentally by David LaPoint.

The expansion is powered exclusively by the electric force

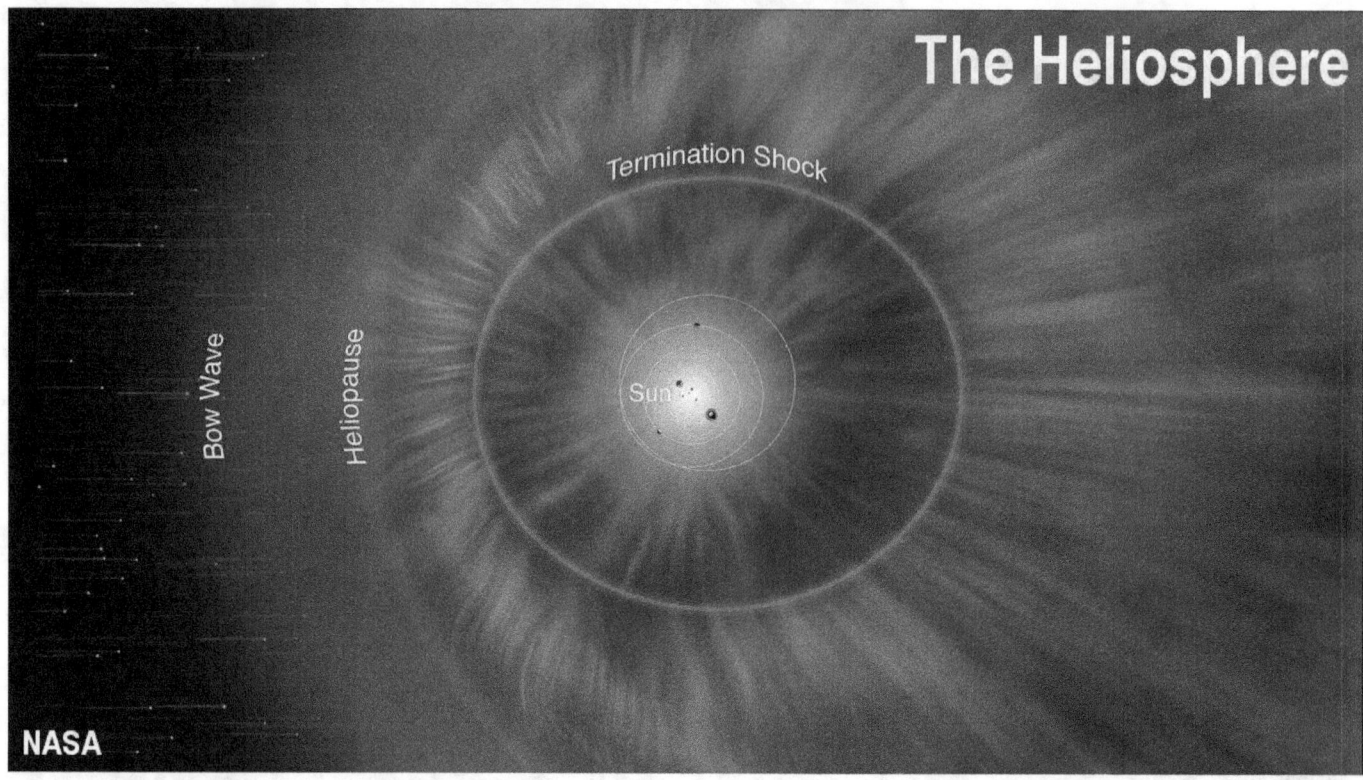

The expansion is powered exclusively by the electric force that accelerates the flowing plasma to a velocity of over 800 kilometers per second and projects it over a distance 100 times the distance between the Sun and the Earth.

Solar wind likened to a heated kettle boiling off steam

The flowing of the solar wind can be likened to a heated kettle boiling off steam. Since water cannot be heated to more than 100 degrees at sea-level pressure, all excess heat that is applied, converts water into steam. Likewise on the plasma sun, when 'excess' plasma pressure bears down on it, some of the excess is vented off as solar wind.

For as long as the solar wind flows

- an example of the amazing solar eclipse photography of Milloslav Druckmueller

This means that for as long as the solar wind flows from our Sun, there is enough plasma pressure surrounding the Sun to keep it operating in a high-powered mode.

When the heat is turned down on a boiling kettle

However, when the heat is turned down on a boiling kettle, steam stops flowing from it and the water cools.

The Sun 'cools' down after the solar wind stops

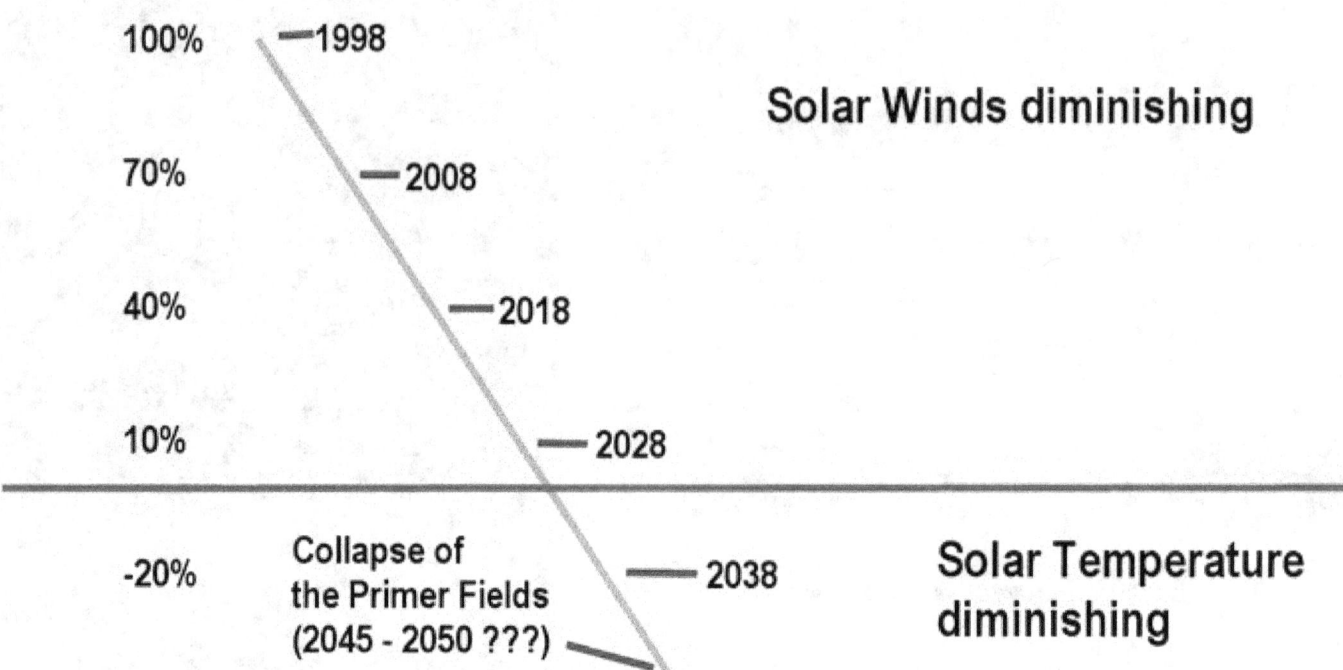

This means that when the plasma pressure around our Sun becomes reduced to the point that the solar wind no longer flows, we find that the fusion reactions are beginning to diminish thereafter, when the solar system continues to weaken. In short, the Sun 'cools' down after the solar wind stops.

Paradox #4, the super-heated solar corona paradox

Also Paradox #4, the super-heated solar corona paradox, is resoled by the plasma sun concept, and in a similar manner than the solar wind paradox is resolved.

Atomic elements in the corona emit light when they are energized electromagnetically. This means, that the more-intensely atomic elements in the corona are energized by the out-flowing solar wind, the more intense will be the light emitted in the corona. The light intensity, in such cases, is measured as temperature.

The temperature in the corona increases by this process with the distance from the Sun. It increases with the distance from the Sun, because the solar-wind velocity increases with the distance from the Sun. As the solar wind flows through the corona, it is accelerated electromagnetically to ever-greater velocity. As the result, several millions of degrees have been measured in the solar corona, in some cases, in the outer regions.

Paradox #5, the mass-density paradox

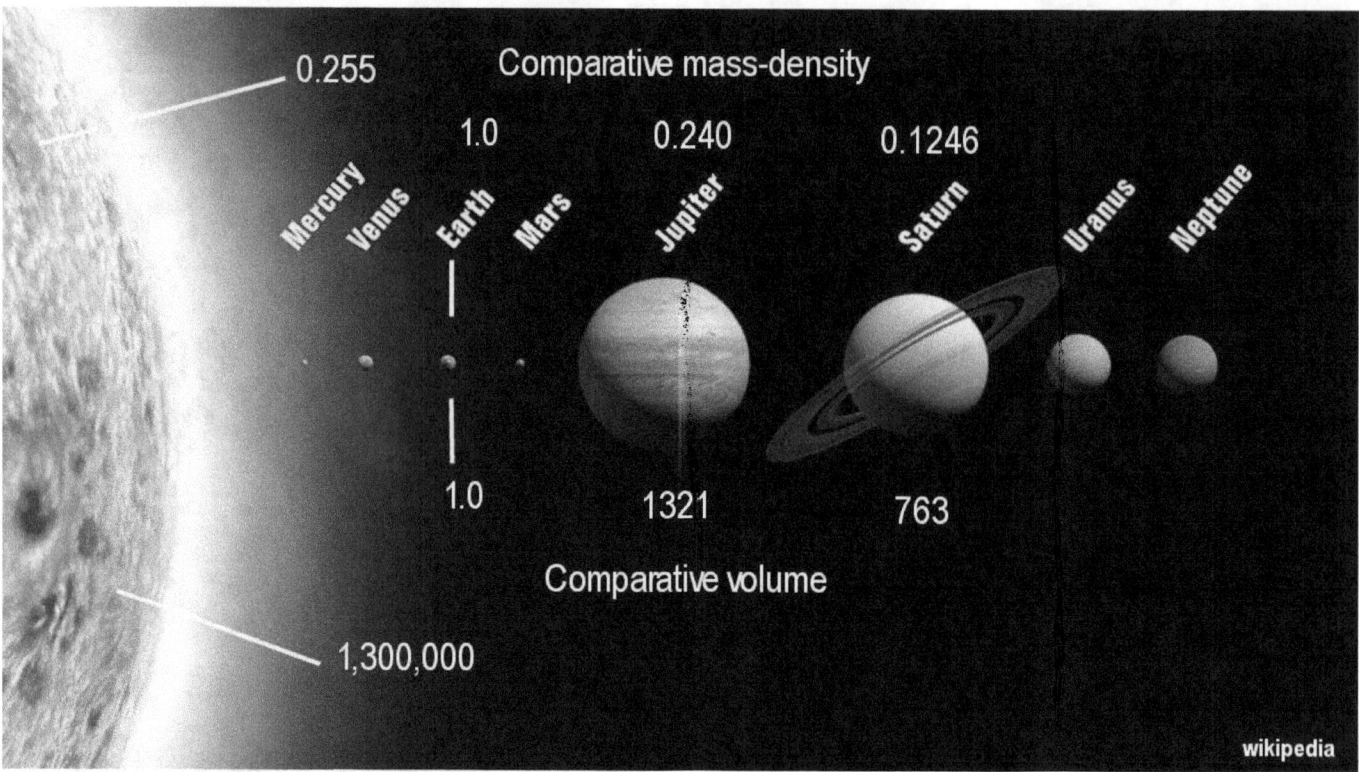

Paradox #5, the mass-density paradox, is of course likewise not a paradox under the plasma Sun.

The extremely low mass-density of the Sun, which is a paradox for the gas-sun, accurately reflects the nature of the plasma sun as a largely empty shell that is electrically 'inflated' from within by the mutually repelling protons in its core. The protons in the core have fewer electrons shielding them from each other.

As an electrically inflated shell of plasma, the known, extremely low mass density of our Sun for its size, is not an enigma, but is inherent in its very nature. Thereby, our Sun's extremely low mass-density, proves our Sun to be a plasma star, and every other star to be a plasma star likewise.

A gas planet is atomic in nature. As such it reflects only gravitational dynamics. These cause extreme gas compression inside a large gas sphere. In contrast, a plasma sun is shaped by electrodynamics where gravity plays only a minor role. The resulting two different structures are incomparable thereby.

The mass-density difference between the Sun and gas planets

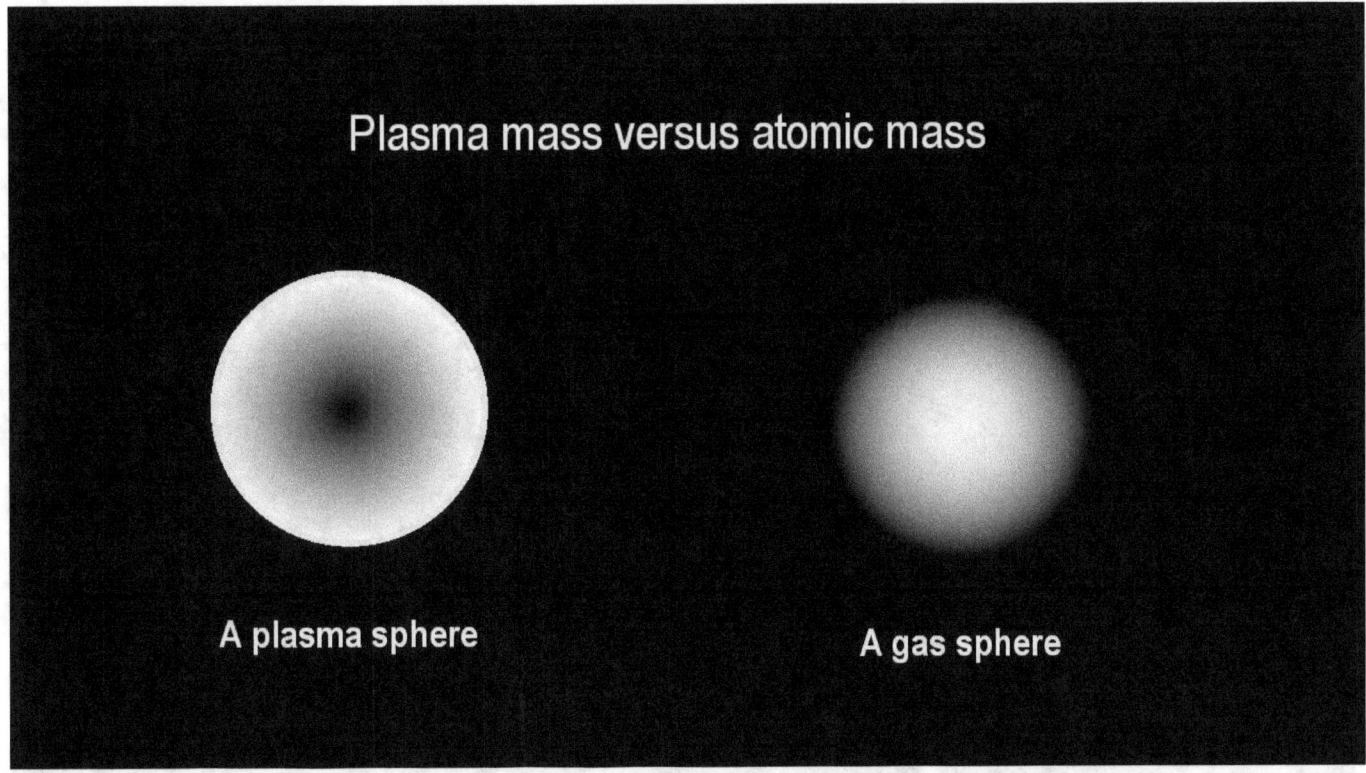

This means that the mass-density difference between the Sun and that of gas planets, is no longer a paradox, because the two are fundamentally different structures by their very nature, with opposite characteristics. A plasma sphere has the least mass-density at its core, where the atomic gas sphere has its greatest mass density at its core, which actually limits the size of a gas sphere.

Paradox #6, the large-star paradox

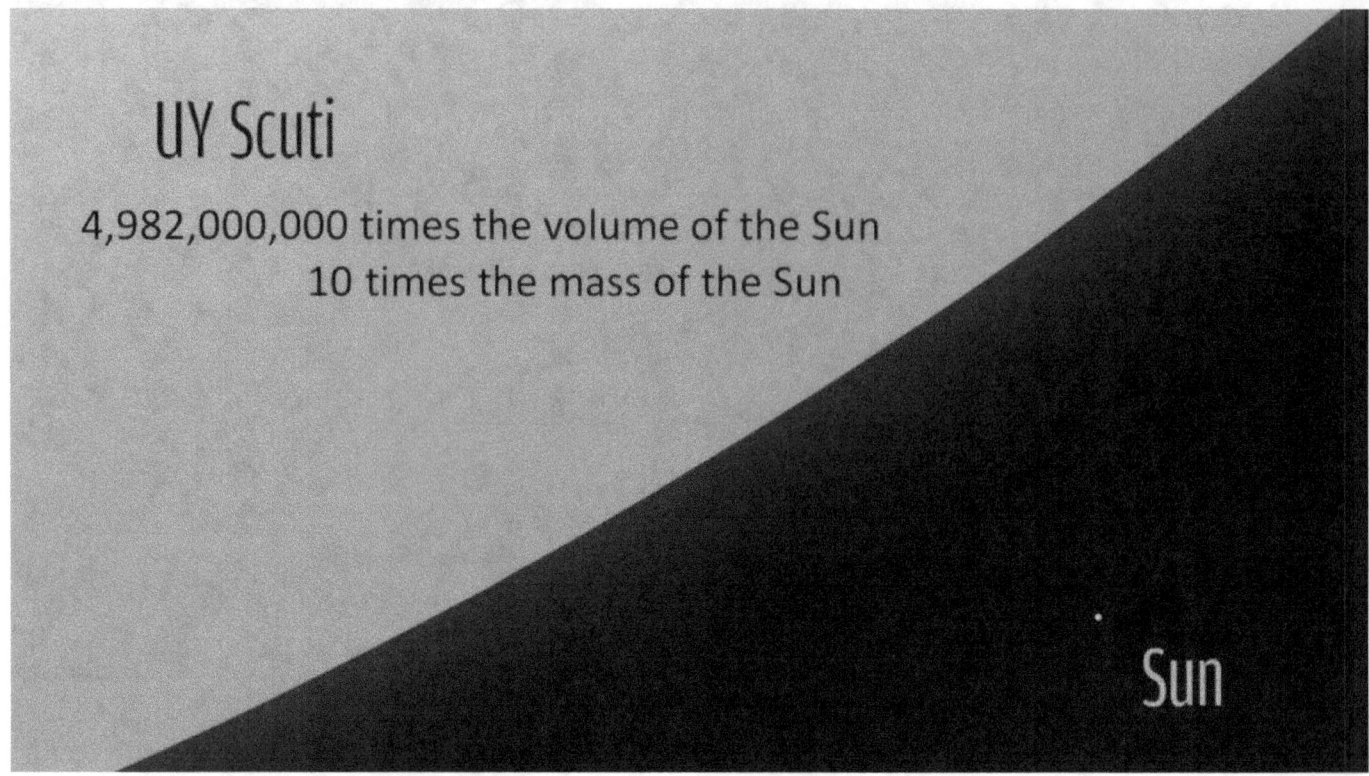

**

The gas-compression limit of a gas sphere is prominently reflected in Paradox #6, the large-star paradox. The enormous size of the star UY Scuti makes it impossible for this star to be a gas star. The gas-compression would cause it explode. But it is totally possible for the star to exist as a plasma star. This means that the large star isn't a paradox then as a plasma star, but proves in its extreme that every sun is a plasma star, as a universal principle.

The star UY Scuti has its ten solar masses spread across its thin plasma shell

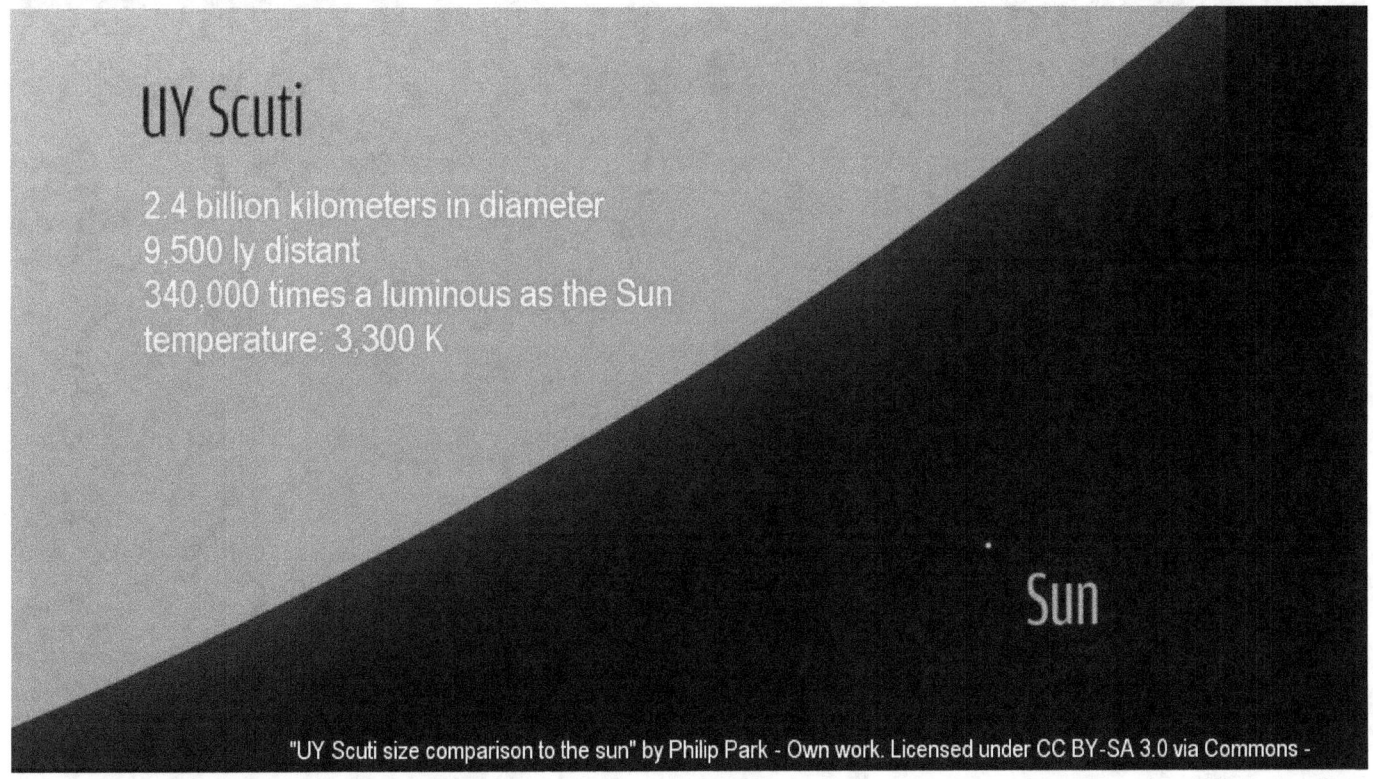

The star UY Scuti has its ten solar masses spread across its thin plasma shell. This means, that since a plasma sun doesn't create the energy it radiates, but merely acts as a catalytic converter of in-flowing energy, the star UY Scuti, which is 1700 times larger in diameter than our Sun, and has a nearly 3 million times larger surface, is able to outshine our Sun 340,000-fold with its mere 10 solar masses. No paradox exists here.

Paradox #7, the cosmic-ray paradox

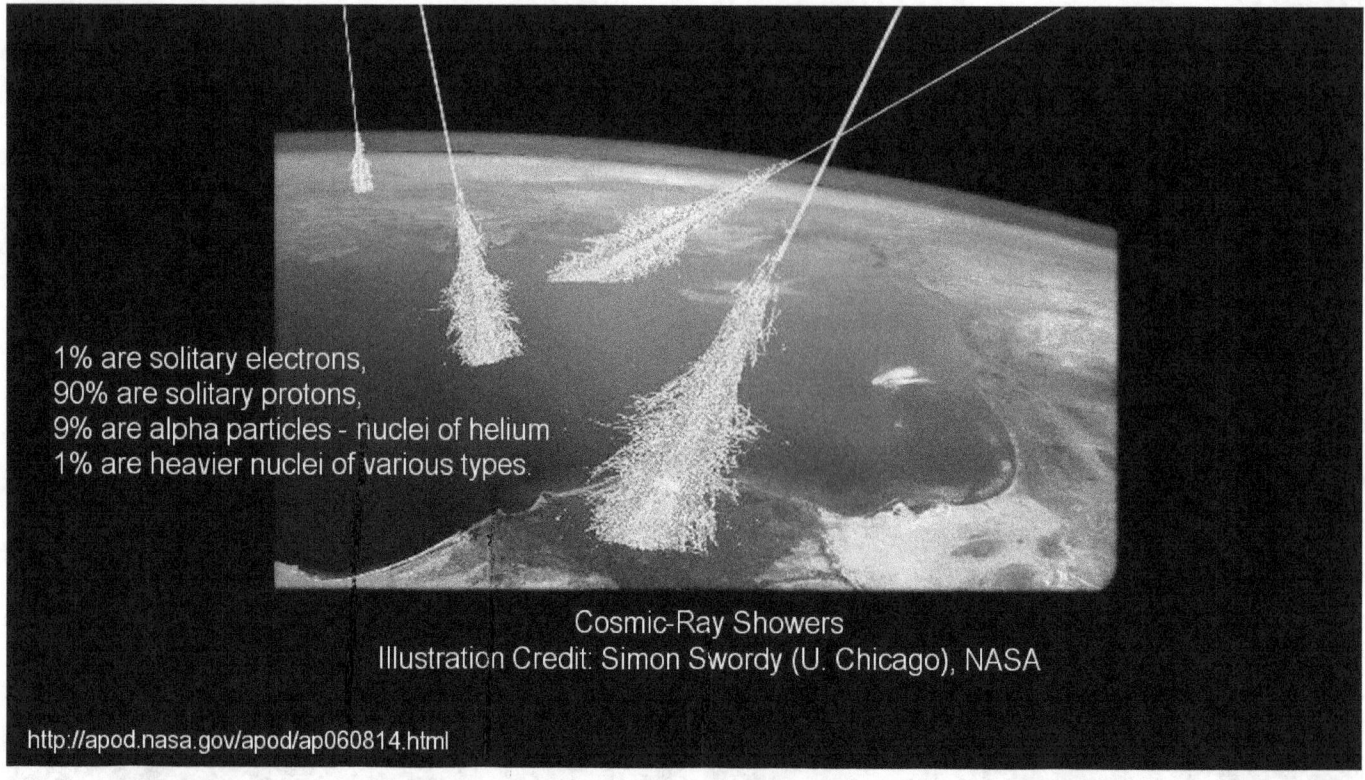

Paradox #7, the cosmic-ray paradox, is likewise non-existent with the plasma sun.

Solar cosmic-ray flux is generated on the surface of the Sun

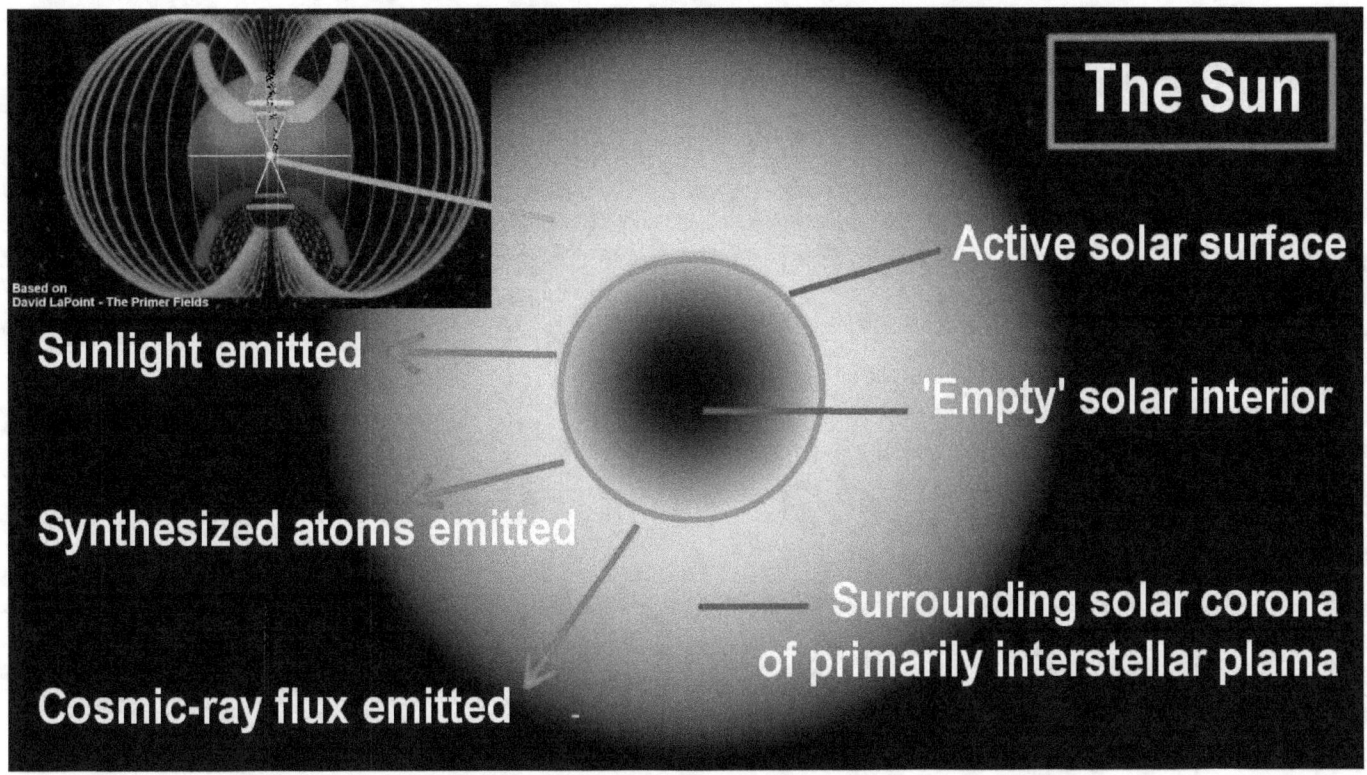

While solar cosmic-ray flux is impossible under the gas-fusion model of mainstream-consensus cosmology, so that all cosmic-ray flux is deemed to be exclusively galactic in nature, solar cosmic-ray flux is as natural for a plasma Sun as is the solar wind.

Solar cosmic-ray flux is natural for a plasma sun, because it is generated on the surface of the Sun. It is as natural as the synthesizing of atomic elements by a plasma sun, and the emission of light and heat by the atomic elements, which all happens on the solar surface. Nothing comes deep within.

Solar cosmic-ray flux is as natural as all that. In fact it is an unavoidable by-product of plasma fusion happening on the surface of the Sun.

A profusion of single plasma particles escaping the fusion cells

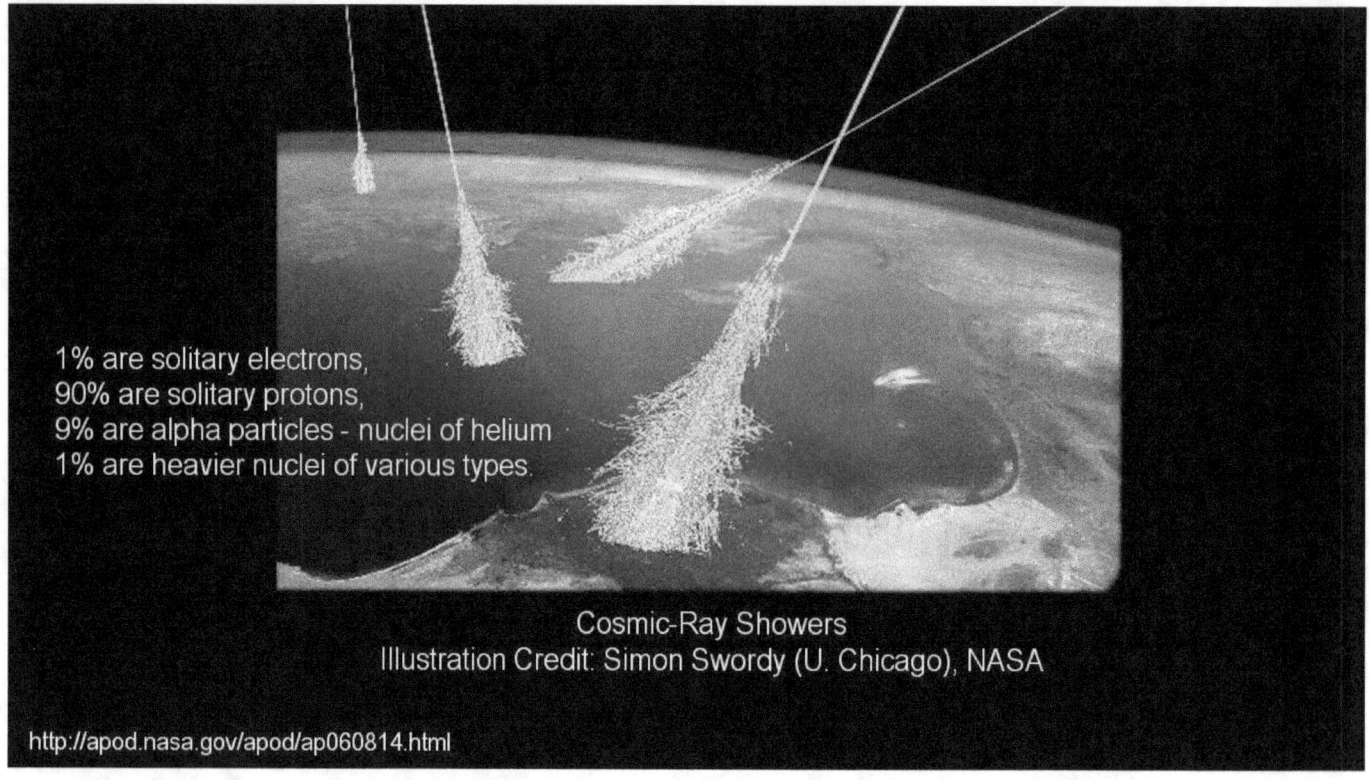

Cosmic-ray flux is a profusion of single plasma particles escaping the fusion cells in highly energized form, in contrast with solar wind that is a 'continuous stream' of low-energized plasma particles.

Cosmic-ray flux originates with the Sun

Reaction products of primary cosmic rays, lifetime and reaction

Tritium (12.3 a): 14N(n, 3H)12C (Spallation)
Beryllium-7 (53.3 d)
Beryllium-10 (1.6E6 a): 14N(n,p a)10Be (Spallation)
Carbon-14 (5730 a): 14N(n, p)14C (Neutron activation)
Sodium-22 (2.6 a)
Sodium-24 (15 h)
Magnesium-28 (20.9 h)
Silicon-31 (2.6 h)
Silicon-32 (101 a)
Phosphorus-32 (14.3 d)
Sulfur-35 (87.5 d)
Sulfur-38 (2.8 h)
Chlorine-34 m (32 min)
Chlorine-36 (3E5 a)
Chlorine-38 (37.2 min)
Chlorine-39 (56 min)
Argon-39 (269 a)
Krypton-85 (10.7 a)

wikipedia - Cosmic Rays

That the vast majority of the cosmic-ray flux that is encountered on Earth, originates with the Sun, is evident in the measured ratios of two types of radioisotopes, Carbon-14 and Berillium-10, which are a part of the reaction products of cosmic-ray interaction with the Earth's atmosphere. Both Carbon-14 and Berillium-10 are exclusively produced by the Sun by cosmic-ray interaction. The rate of their production is measurable.

That most of Carbon-14 is caused by the Sun

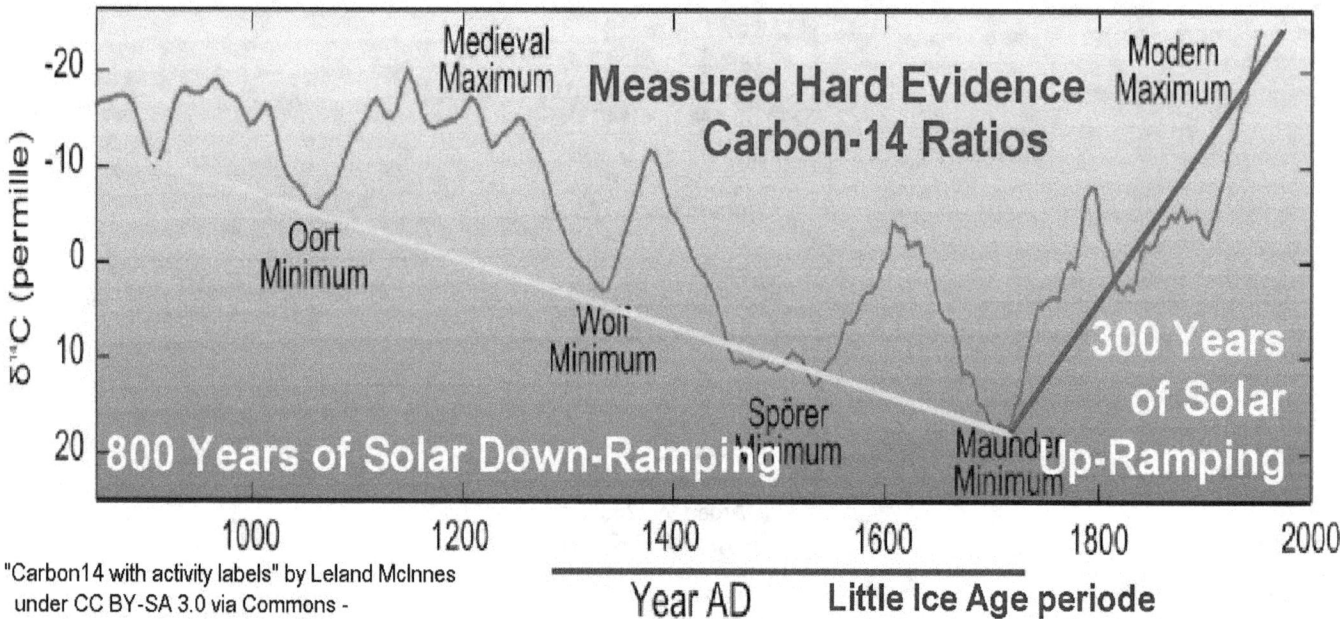

"Carbon14 with activity labels" by Leland McInnes under CC BY-SA 3.0 via Commons -

Carbon-14 is also exclusively produced by cosmic-ray flux. That most of Carbon-14 is caused by the Sun is evident by the direct relationship of the climate on Earth with the changing Carbon-14 ratios. When the solar activity is strong, the ratio is low, and the climate is warm. The changing ratio proves that the global warming from the 1700s onward, was caused exclusively by the Sun being a plasma star that emits cosmic-ray flux in proportion of solar activity.

When the solar activity is strong

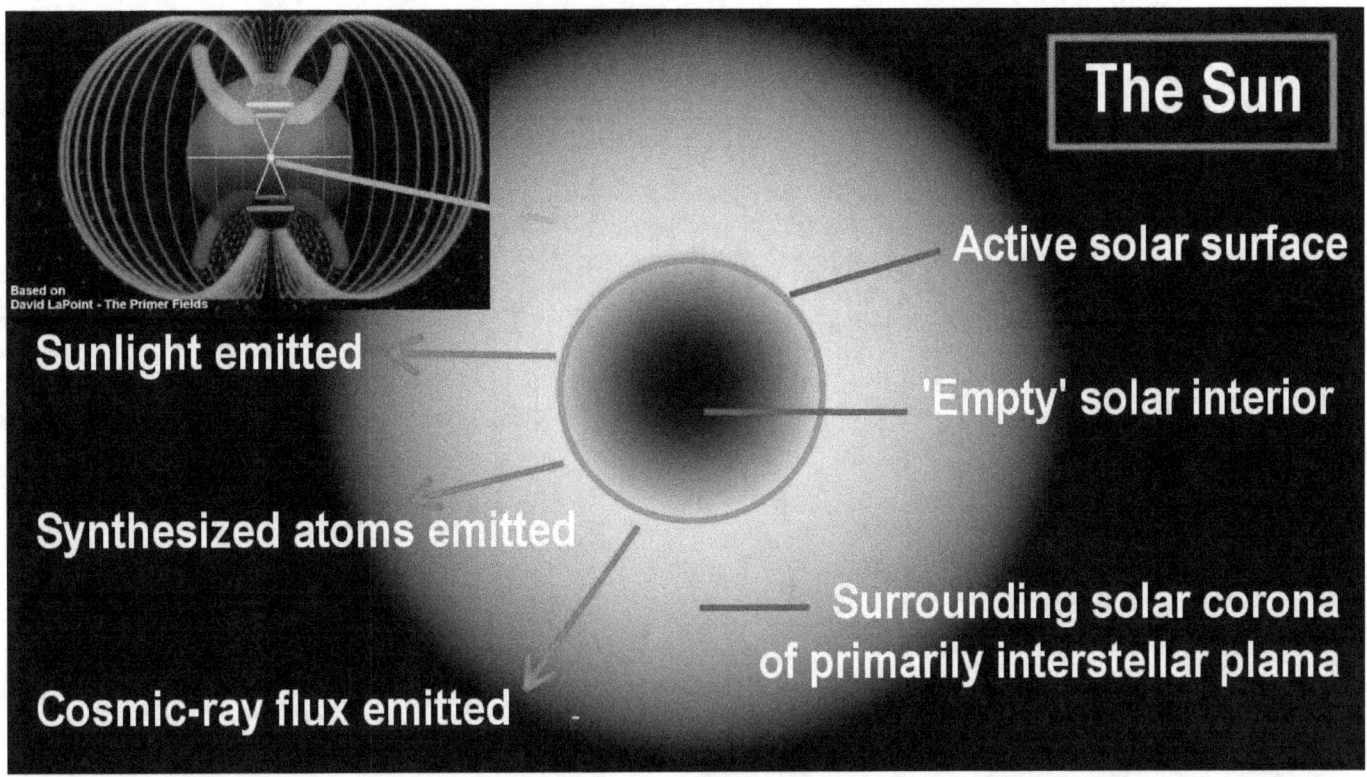

When the solar activity is strong, the plasma sphere that surrounds the plasma-star Sun is likewise strong, whereby greater volumes of cosmic-ray particles become trapped in it, and lesser volumes affect the Earth. Inversely, in times when the solar activity is weak, a lesser sphere of plasma surrounds the Sun, which causes the Sun to be less active, while larger volumes of solar cosmic-ray events penetrate the lesser plasma sphere and affect the Earth.

That the fluctuating cosmic-ray flux is solar in origin

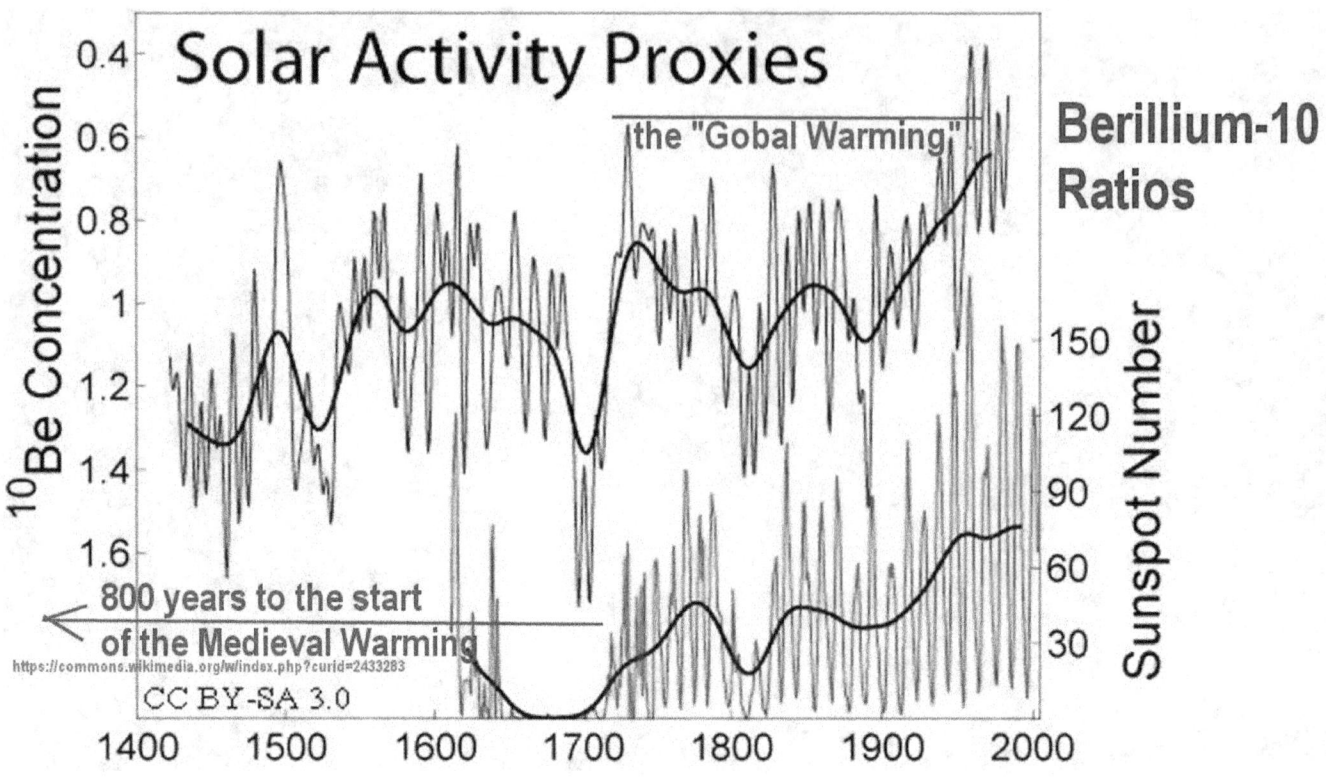

That the fluctuating cosmic-ray flux is solar in origin is evident by the fact that the isotope ratios follow both the solar cycles and the known climate fluctuations.

A third proxy for solar cosmic-ray flux

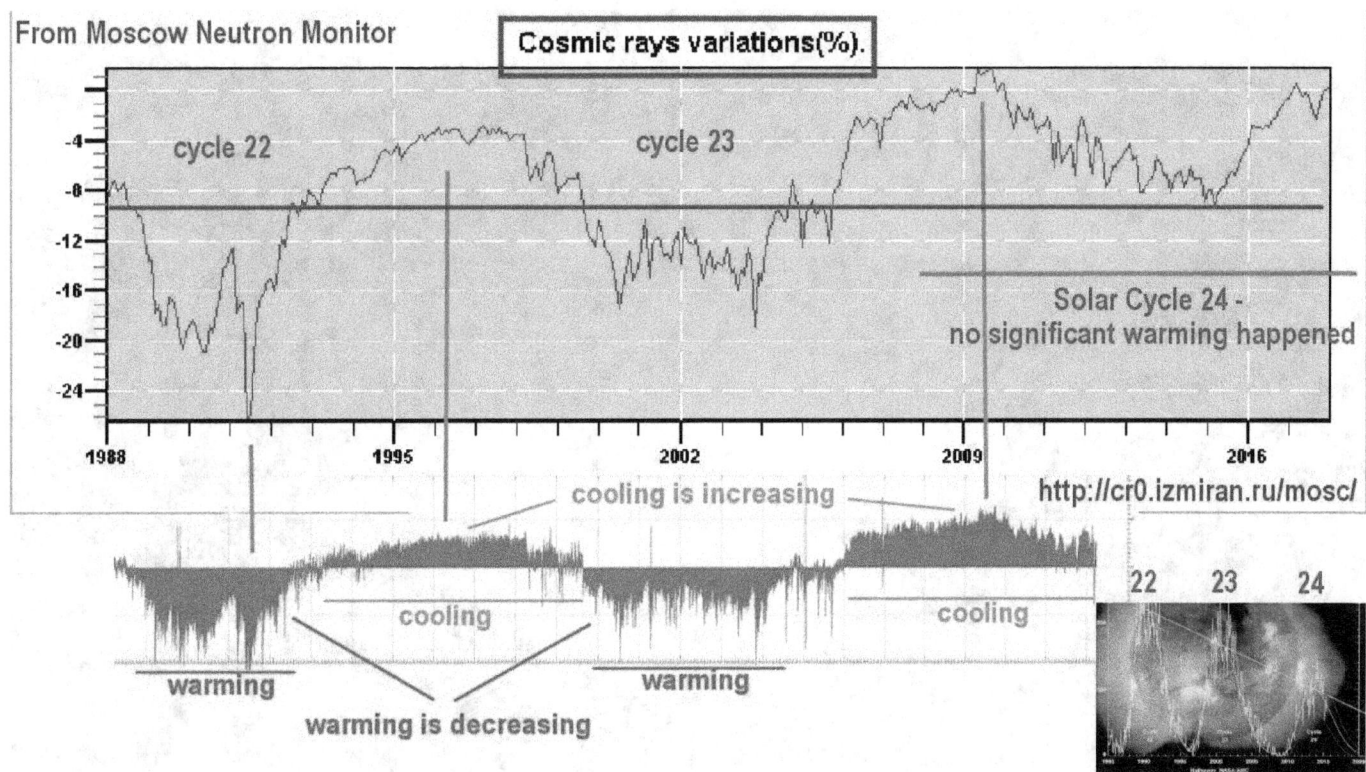

In more recent times a third proxy for solar cosmic-ray flux is being measured, by measuring the atmospheric neutron flux density. Since cosmic-ray interaction also generates free neutrons in the atmosphere, which are short-lived, the measuring of the neutron density provides a near perfect means to measure solar cosmic-ray flux in real time, which of course follows the solar cycles in perfect synchronism. The synchronism proves that cosmic-ray flux is primarily solar in origin, and that the Sun is a plasma star that is exclusively able to emit cosmic-ray events.

A gas star is cosmic-ray dead

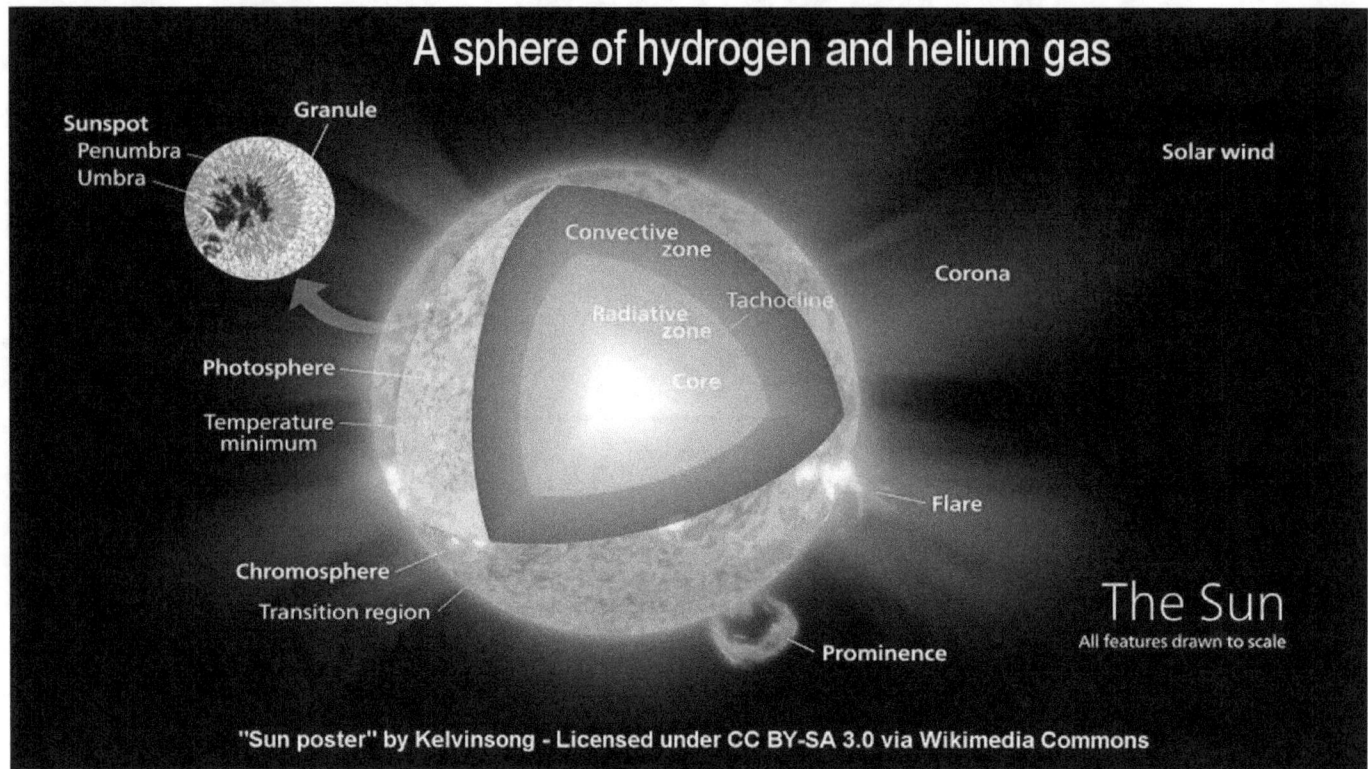

A gas star is cosmic-ray dead, because cosmic rays cannot penetrate the half-million kilometer thick gas shell that is deemed to surround the core of the theorized gas sun. This fact renders the gas-sun theory false, according to measured evidence.

Paradox #8 of the 11-year solar cycles

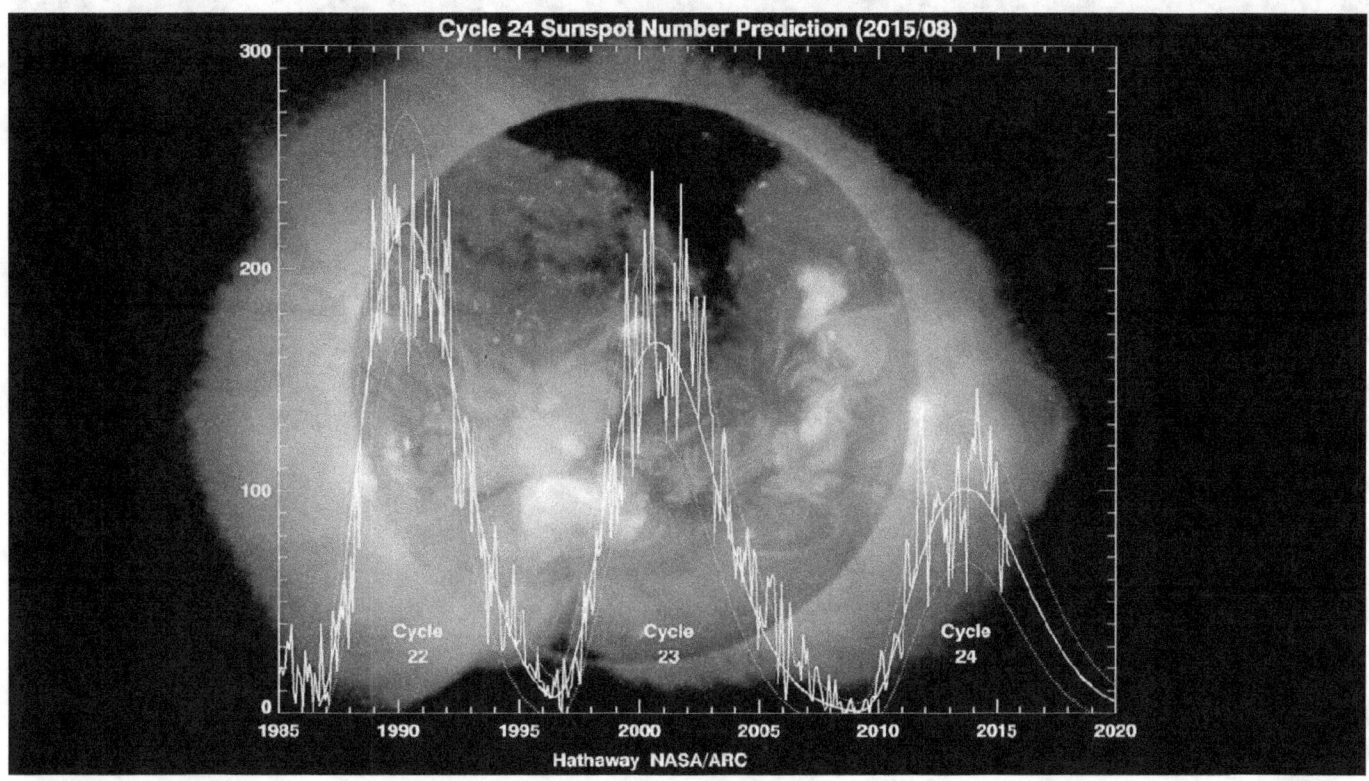

Paradox #8 is a similar paradox. It is the paradox of the 11-year solar cycles and their changing intensity that shouldn't be possible for a gas sun, but are happening.

The gas-sun model doesn't support solar cycles

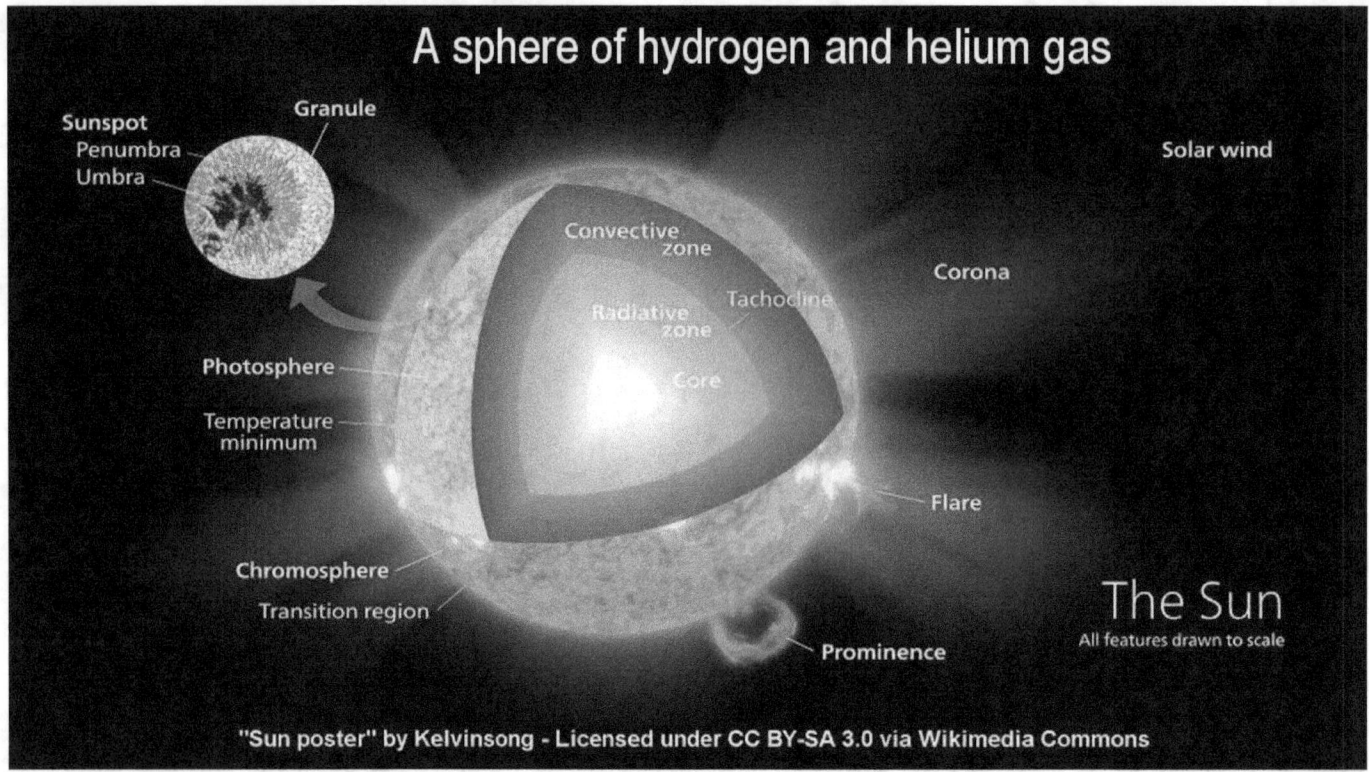

The gas-sun model doesn't support the phenomenon of fast changing solar cycles, except in dreaming. The gas-sun model that is deemed to have nuclear fusion heat generated at its core, which supposedly oozes to the solar surface over a span of 10,00 years to 170,000 years, potentially up to 30-million years as has also been proposed, lacks the capacity to generate the short 11-year solar cycles that are a standard feature of the Sun. That's the paradox.

The solar cycles are not caused by the Sun itself

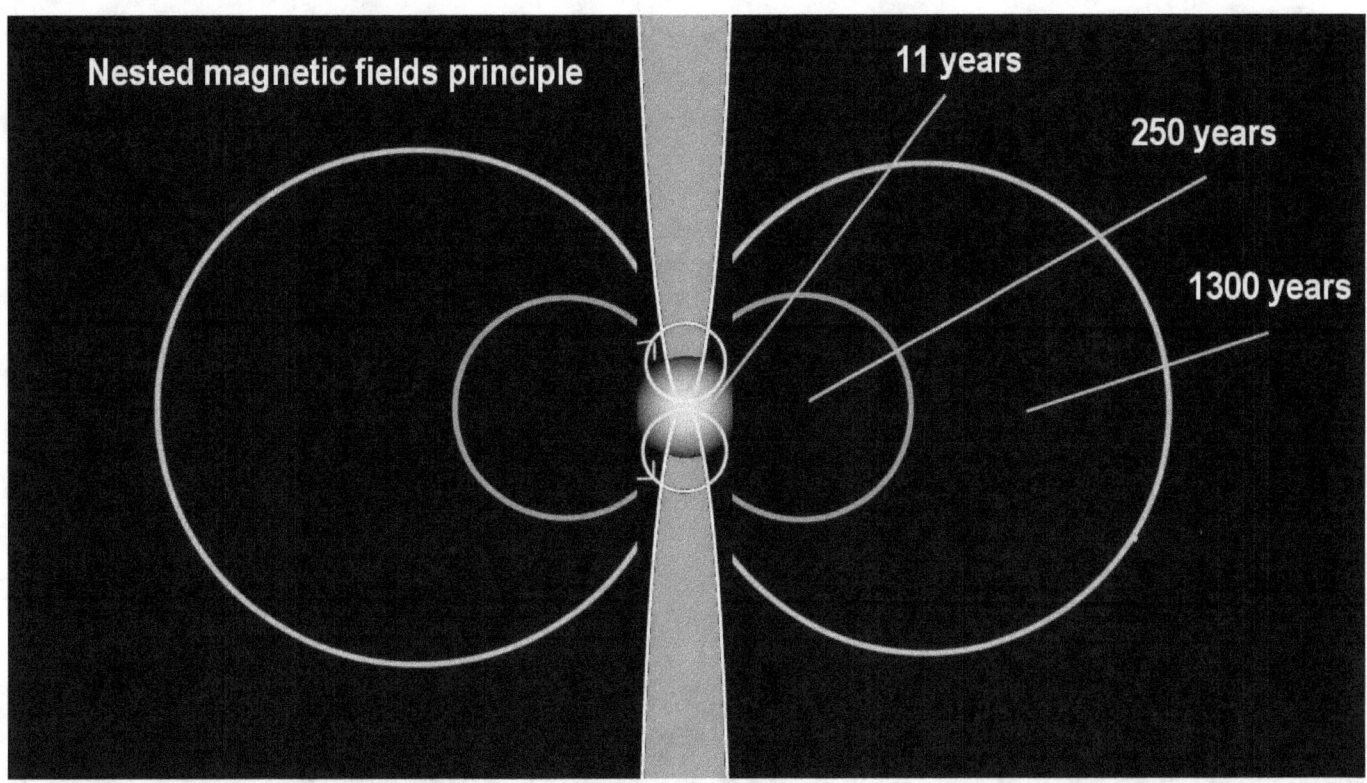

The plasma-sun model is free of this paradox, because the solar cycles and all the numerous other cycles that affect our climate, are not caused by the Sun itself, but are caused by resonating plasma features external to the Sun, within the solar system, but which affect the Sun.

Paradox #9 is located in the sunlight

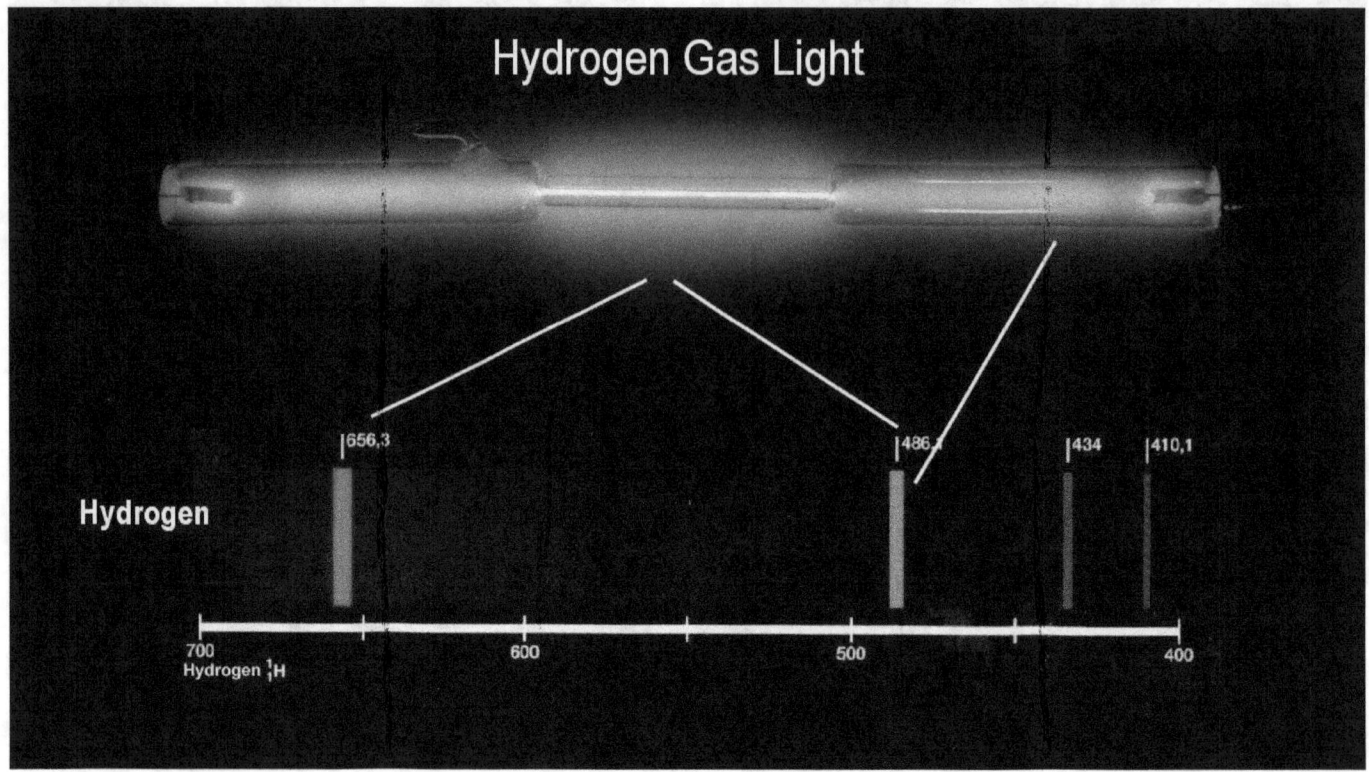

Paradox #9 is located in the sunlight that the Sun emits. If the Sun was a gas sphere of mainly hydrogen, then the sunlight would be limited to the sparse colors of the hydrogen gas emission spectrum.

Homogenous spectrum of colors in the sunlight

Since we see a vastly richer, wider, and homogenous spectrum of colors in the sunlight, a paradox unfolds here that we don't encounter with the plasma Sun, though many would disagree. The sunlight paradox is best resolved by solving Paradox #10.

Paradox #10 is the paradox of the profusion of atomic elements

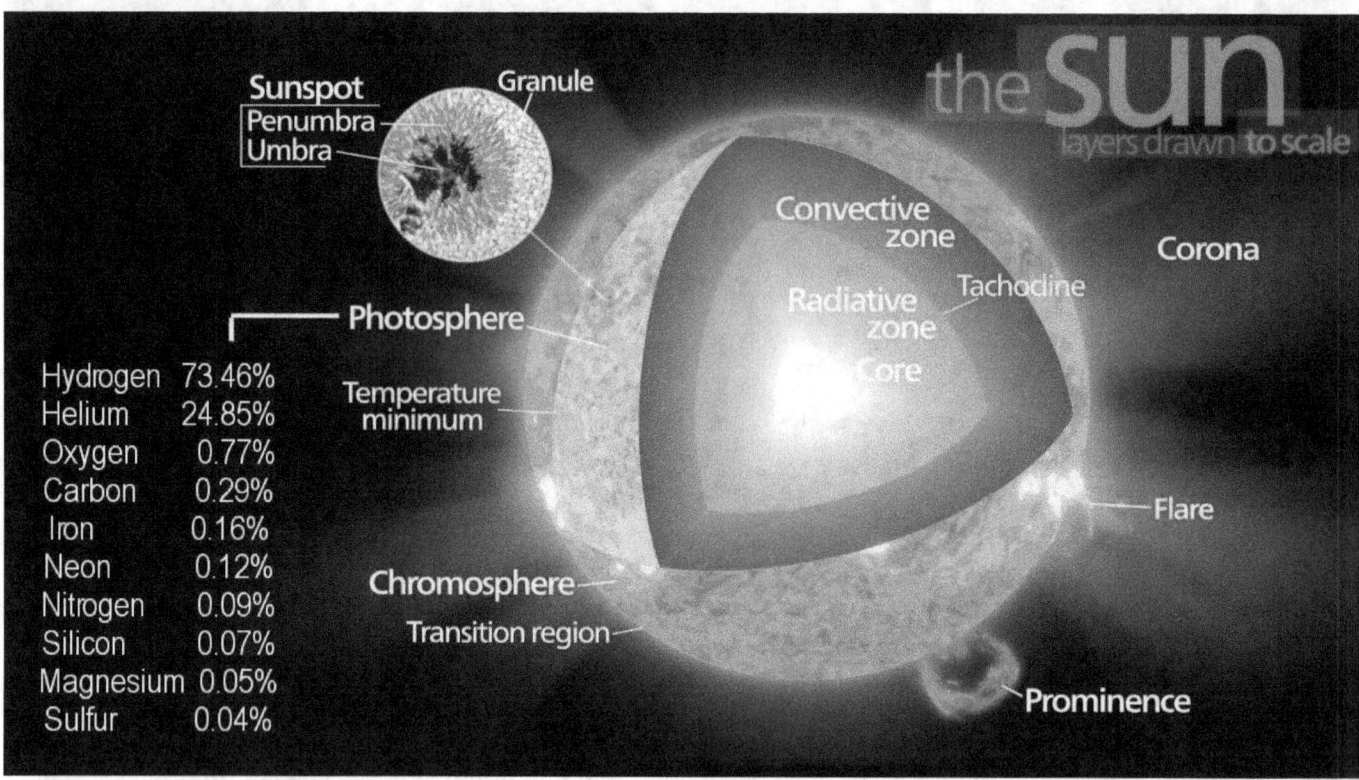

Paradox #10 is the paradox of the profusion of atomic elements that have been detected on the solar surface. If the Sun would draw these atomic elements from surrounding space, they would not dwell on the surface, but would be drawn to the Sun's center of gravity, to the core.

The plasma Sun synthesizes all atomic elements right on its surface

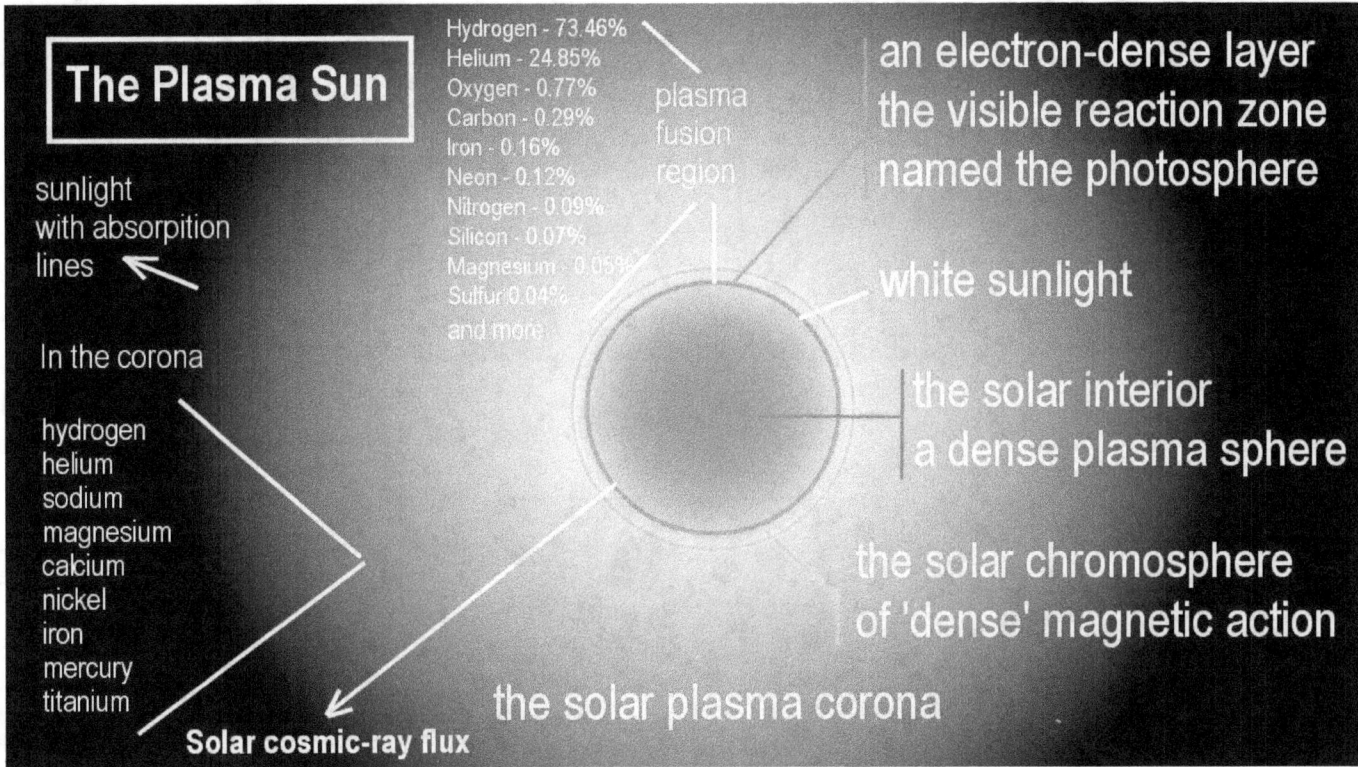

The plasma Sun is free of this contradiction, because the plasma Sun synthesizes all atomic elements that naturally exist, right on its surface, by plasma fusion. This means that we have a wide range of atomic elements on the solar surface where they are created, all emitting light in different spectra, so that the combination yields the wonderfully wide band of color that is contained in the sunlight.

Only the plasma Sun can generate the rich spectrum of the sunlight

This means that only the plasma Sun can generate the rich spectrum of the sunlight that we see, whereby the paradox is solved.

Paradox #11 of the differential rotation of the Sun

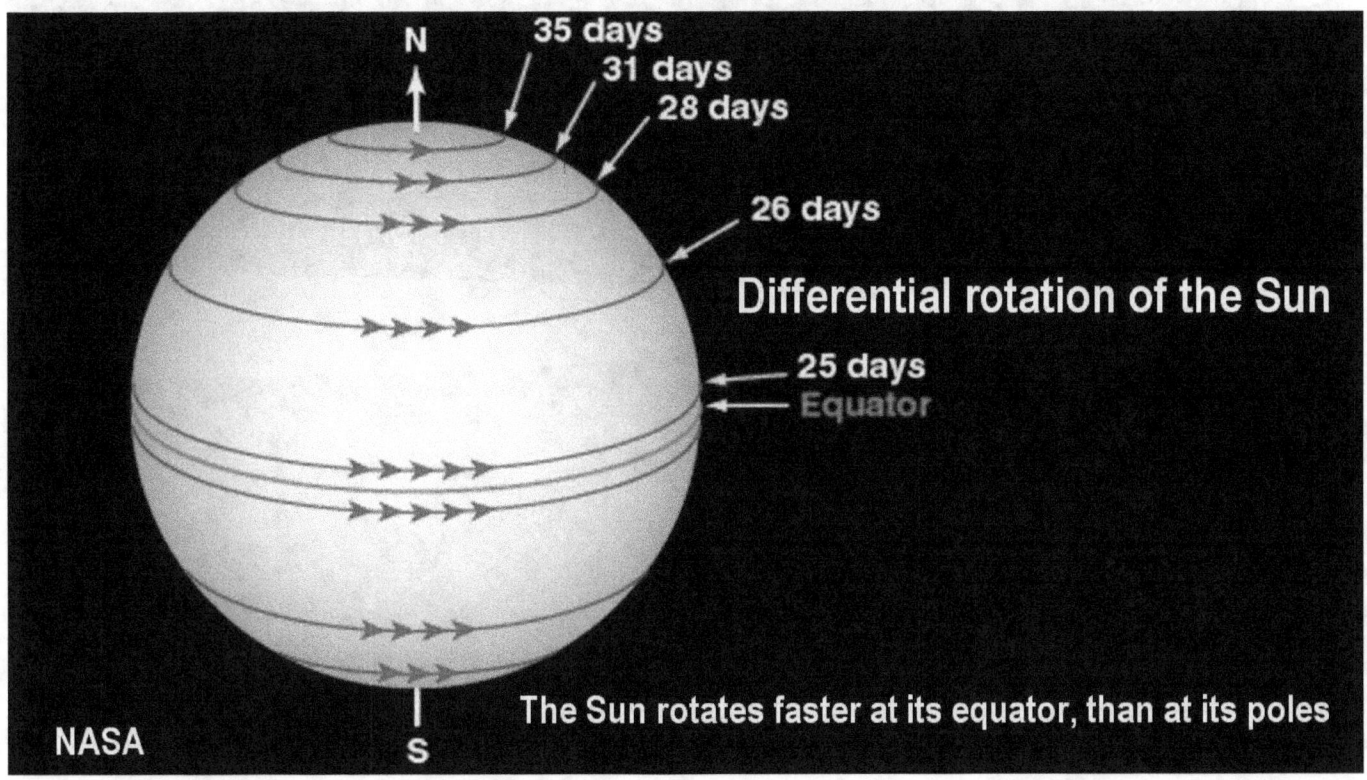

Paradox #11 is the paradox of the differential rotation of the Sun, which is not inherent in the dynamics of a gas sphere.

It is a feature of the plasma Sun that is affected externally by the operating features of the larger solar system in which the Sun is located.

The faster rotation of the Sun at its equator, indicates that the Sun's rotation is externally powered by rotating plasma streams within the larger solar system.

A plasma Sun operates by drawing interstellar plasma streams

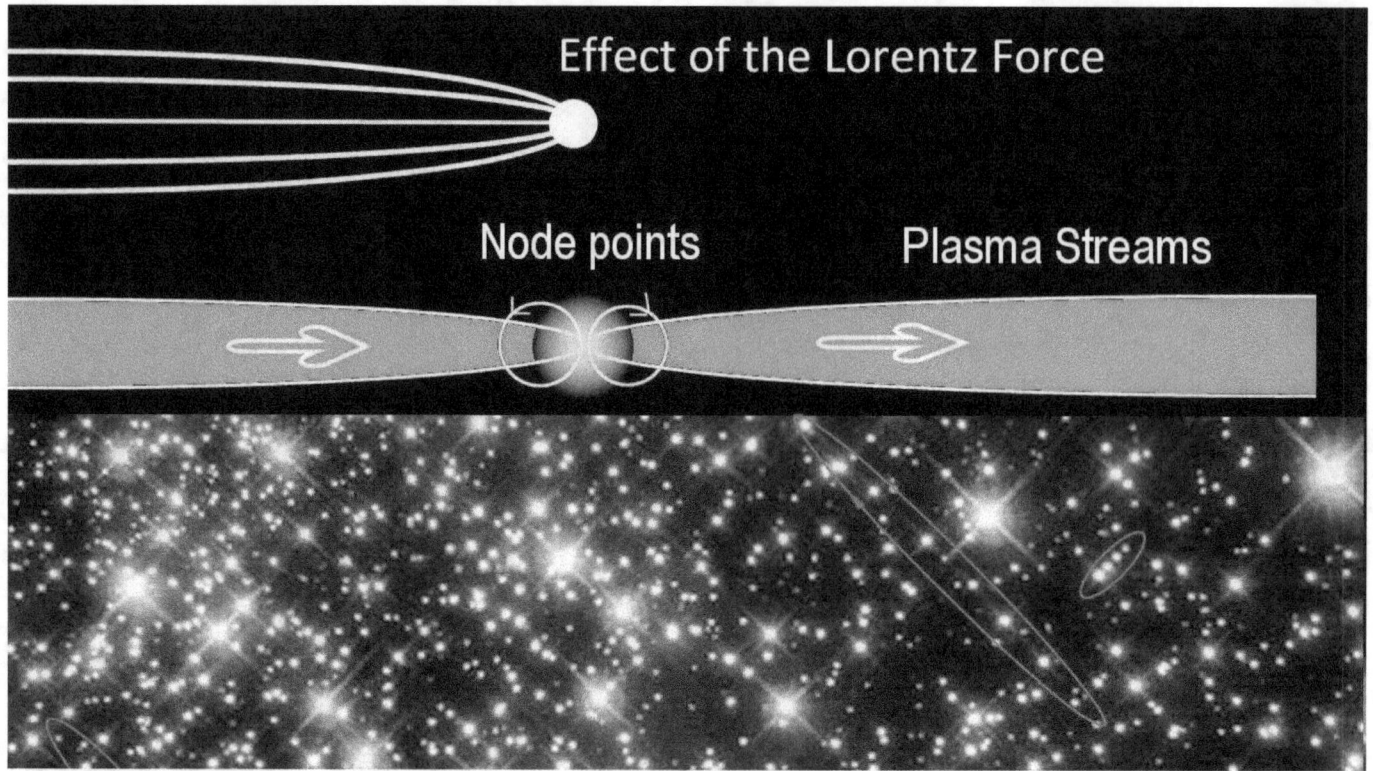

A plasma Sun operates by drawing interstellar plasma streams to it, with which it interacts, by which it is powered.

However, plasma particles carry an electric charge, which when set in motion generate a magnetic field around them. The combined magnetic fields draws the moving plasma together into a self-confined stream, by what is termed the Lorentz Force. The increasing density in the stream, of course, increases the magnetic field that pinches the stream together ever tighter.

When the magnetic fields tangle up

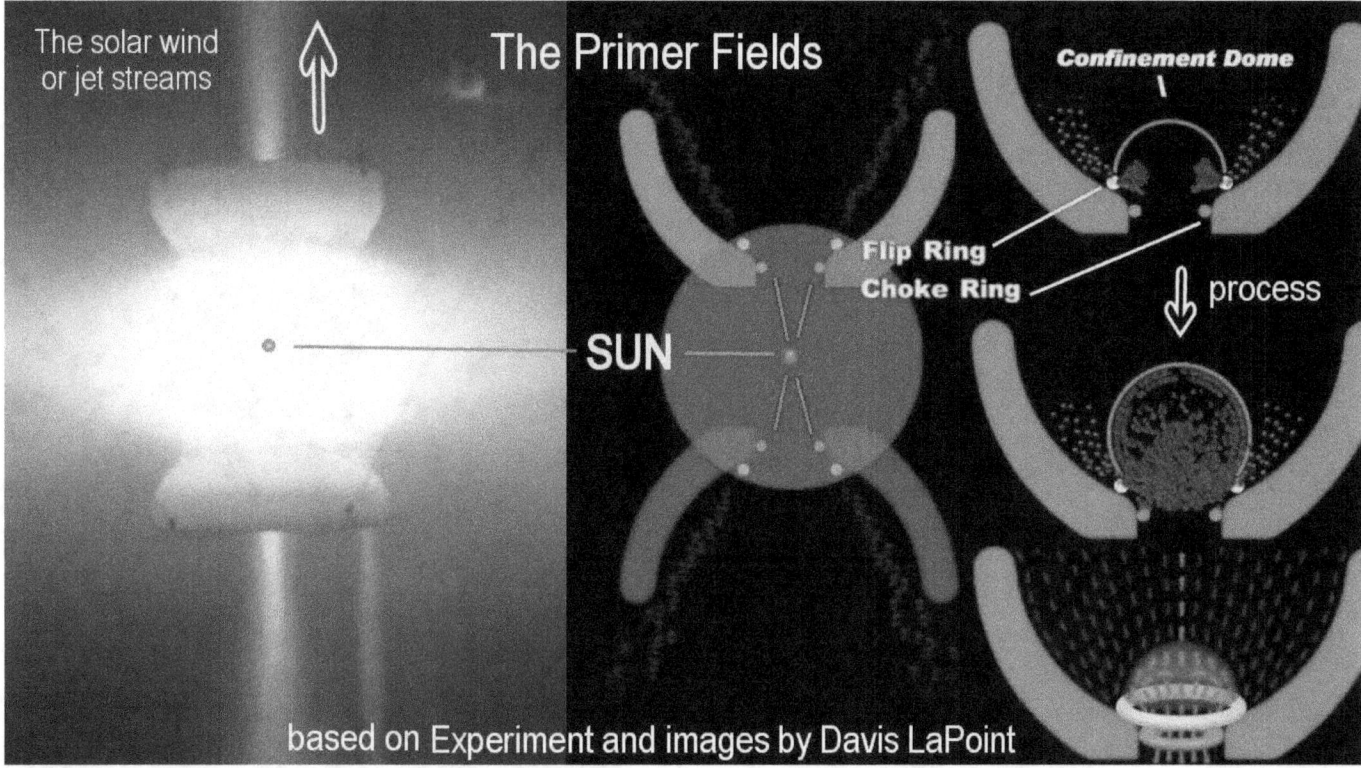

By this increase a point is reached when the magnetic fields tangle up, by which the plasma begins to flow backwards. The back-flow, however, is limited by a curled back magnetic field that keeps the plasma confined and thereby evermore self-concentrating.

The plasma stream surrounds the Sun

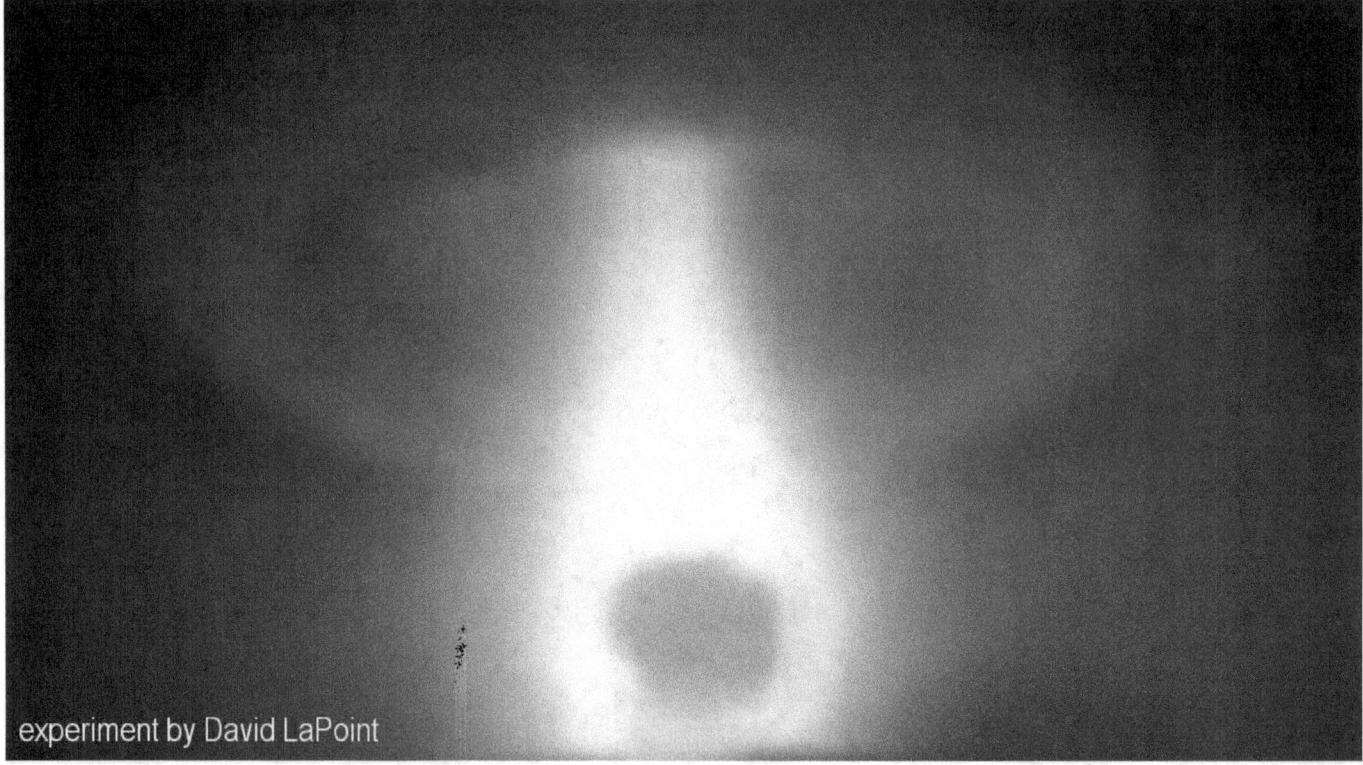

By this process a point is reached by which a focused stream of concentrated plasma flows out from the confinement cell unto a plasma sun, which every Sun is. The plasma stream surrounds the Sun.

An outer ring of 56 plasma filaments with rotating magnetic fields

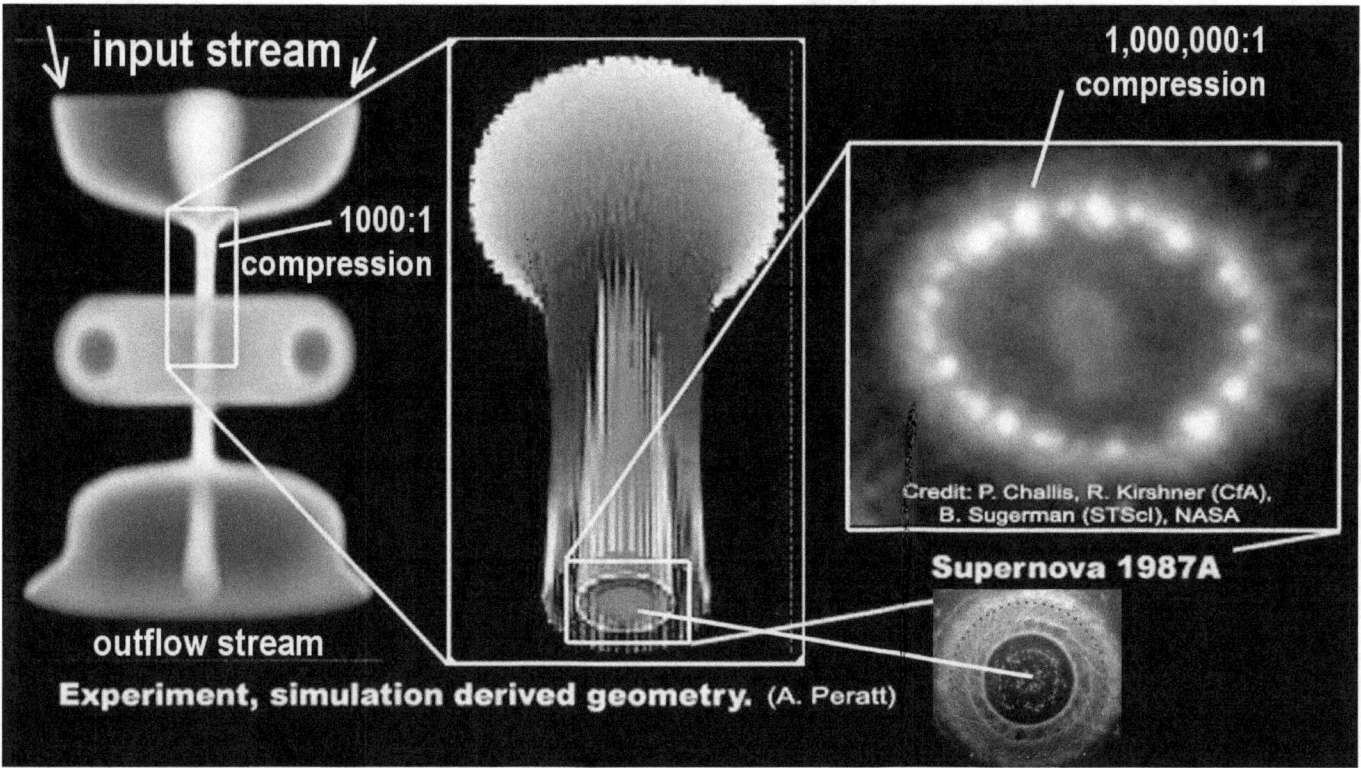

High-power plasma discharge experiments have revealed that that the focused plasma stream that flows out of the confinement structure is made up of an inner core, and an outer ring of 56 plasma filaments with rotating magnetic fields, which is focused onto a sun.

The Sun becomes surrounded by rotating magnetic fields

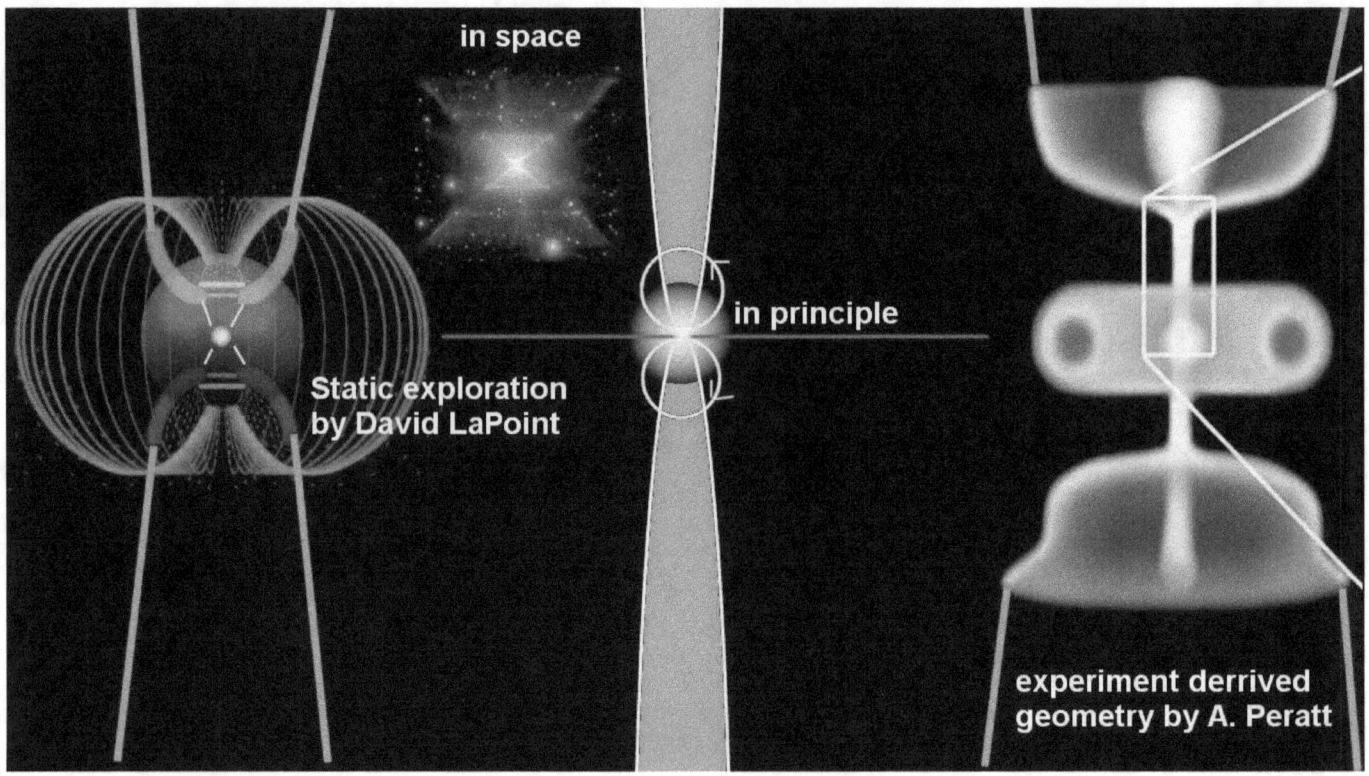

The Sun becomes surrounded thereby, not only by plasma, but also by rotating magnetic fields.

The Sun's differential rotation proves that the Sun is a plasma star

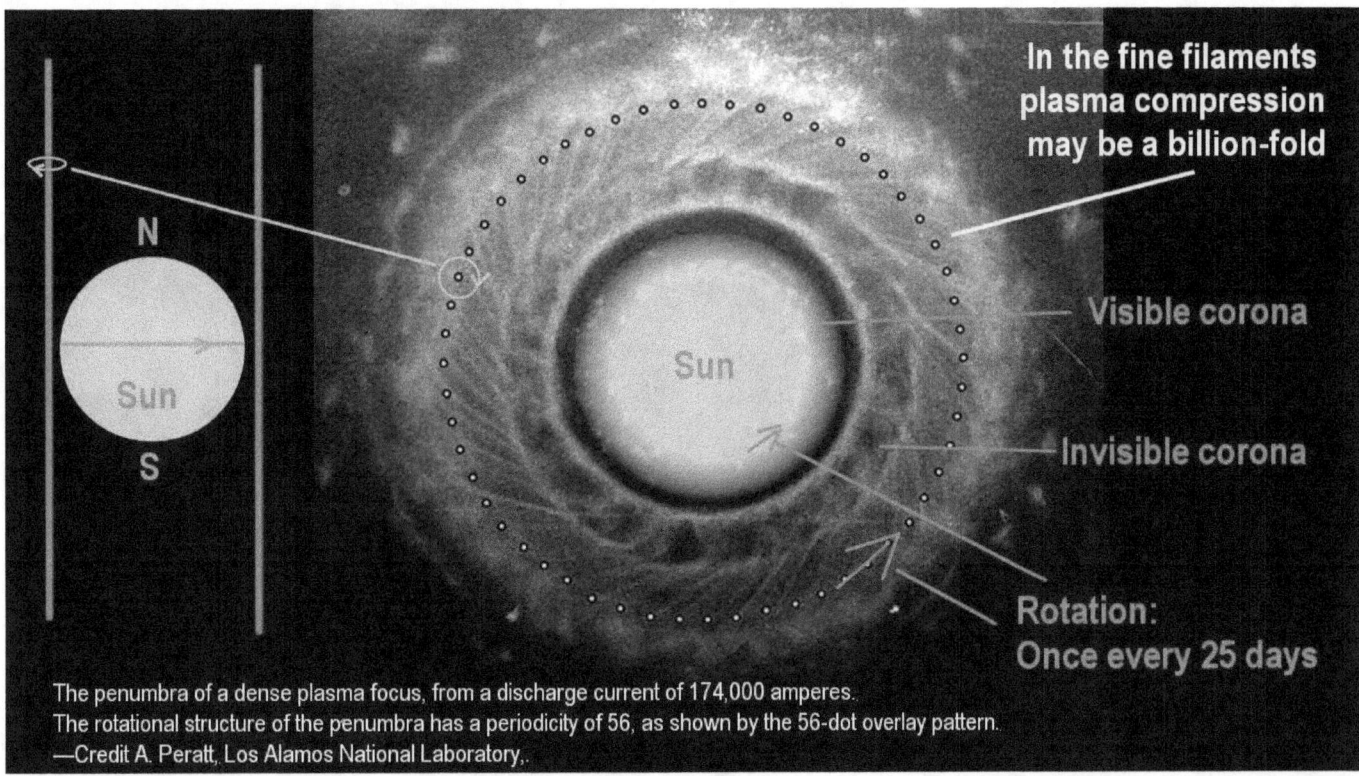

The penumbra of a dense plasma focus, from a discharge current of 174,000 amperes. The rotational structure of the penumbra has a periodicity of 56, as shown by the 56-dot overlay pattern.
—Credit A. Peratt, Los Alamos National Laboratory,.

The rotating magnetic fields, naturally affect the surface plasma of the Sun. The induced rotation is strongest at the Sun's equator, and weakest at the more distant poles. Thus, the Sun's differential rotation proves that the Sun is a plasma star, operating within a complex plasma solar system.

Solar operation consumes a portion of the in-flowing plasma

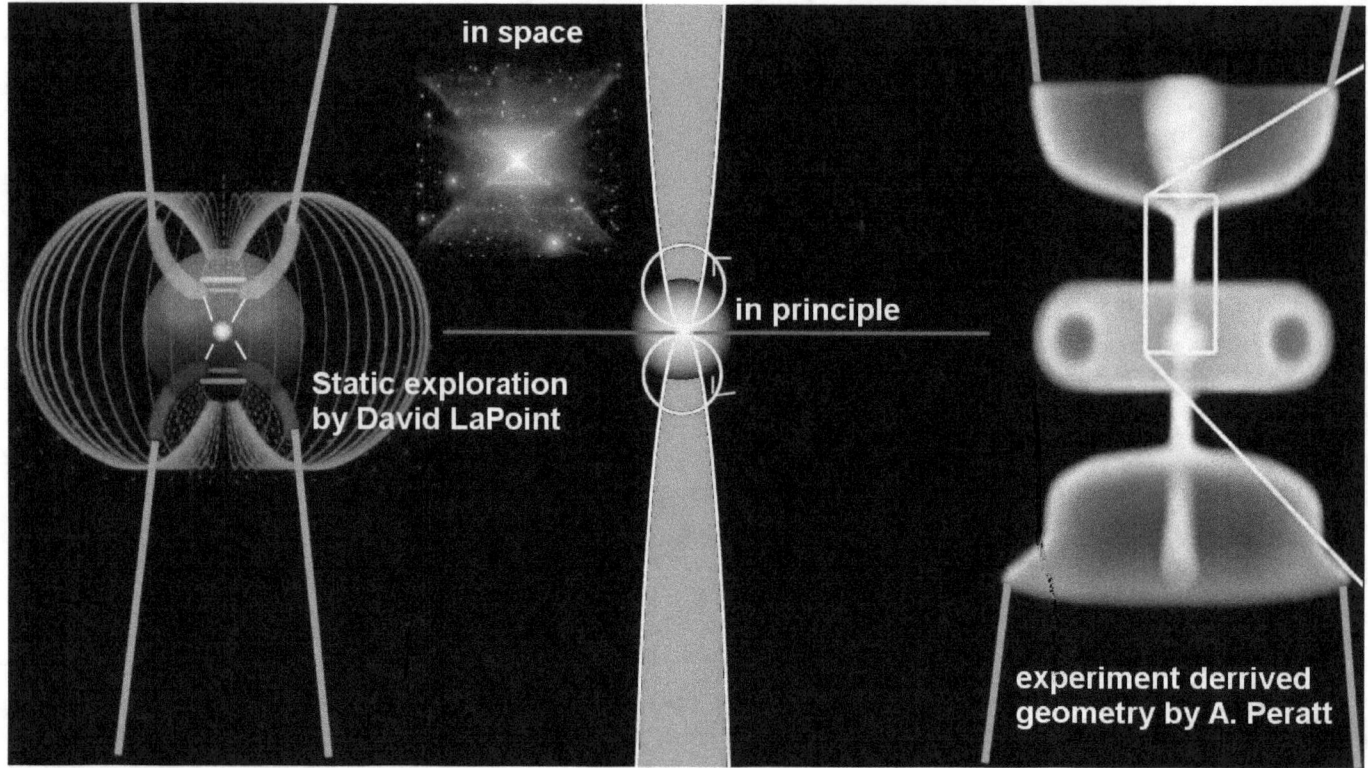

The entire system is driven by magnetically self-focused interstellar plasma streams, that all reflect the basic principles of primer fields in operation. The plasma streams flow into the solar system across long distances of interstellar space. When the plasma is eventually focused onto a plasma sun, typically by magnetic primer fields, the thereby facilitated solar operation consumes a portion of the in-flowing plasma with a process of atomic synthesis.

All the atoms that exist, were synthesized on the surface of a Sun

All the atoms that exist, were synthesized on the surface of a Sun by consuming plasma.

The consumption of plasma creates the sink effect that enables plasma to flow

The consumption of plasma with atomic syntheses is critical for the entire solar process to function. It creates the sink effect that enables interstellar plasma to flow. When plasma becomes bound up into atoms, which happens in a balanced fashion, the resulting construct becomes electrically neutral. The, thereby bound-up plasma particles, functionally disappear from the electrodynamics landscape, as if they had never existed. They flow away into space.

The fusion of plasma into atoms creates that vital sink effect without which nothing would flow. It creates a kind of destination for the plasma streams to flow to, and flow into, like a bottomless pit. The bottomless pit keeps the entire system of plasma streams in motion. The solar system thereby acts as a sink for the galactic system and enables its flow. Without the plasma sun that consumes plasma in rich measures, by binding it up into atoms, the universe would likely not exist. Nothing would flow. Nothing would be created.

All the atoms for the planets were synthesized by the Sun

All the atoms for the planets were synthesized by the Sun

The plasma sun thereby comes to light as the heart that keeps the blood flowing.

Whatever portion is not consumed, simply flows on

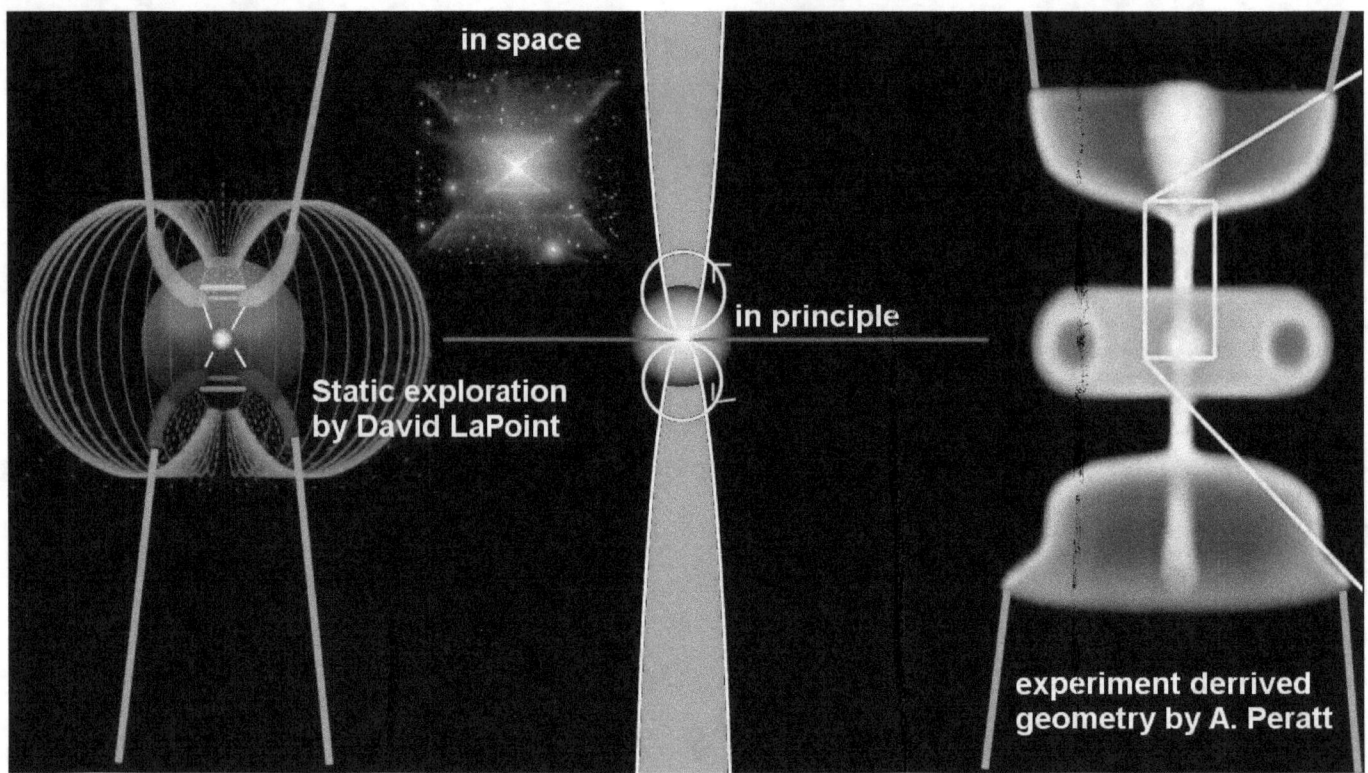

Whatever portion of the interstellar plasma that flows into the solar system, is not consumed, simply flows on to the next star. It flows on as a weaker plasma stream and expands again through a complimentary set of magnetic fields that operate in reverse and expand the excess plasma flow into an out-flowing interstellar plasma stream.

That's why the stellar primer fields are complimentary in nature

That's why the stellar primer fields are complimentary in nature.

The out-flowing stream subsequently gathers up plasma

The out-flowing stream subsequently gathers up plasma from galactic space as it flows, on its path to another star nearby.

The term, 'the Primer Fields'

David LaPoint - The Primer Fields

The term, 'the Primer Fields' was likely pioneered by the researcher David LaPoint who had conducted extensive static experiments of the magnetic structures to discover the underlying principles. He published several videos about it, under the title "The Primer Fields."

Stars are often lined up into long strings of stars in cosmic space

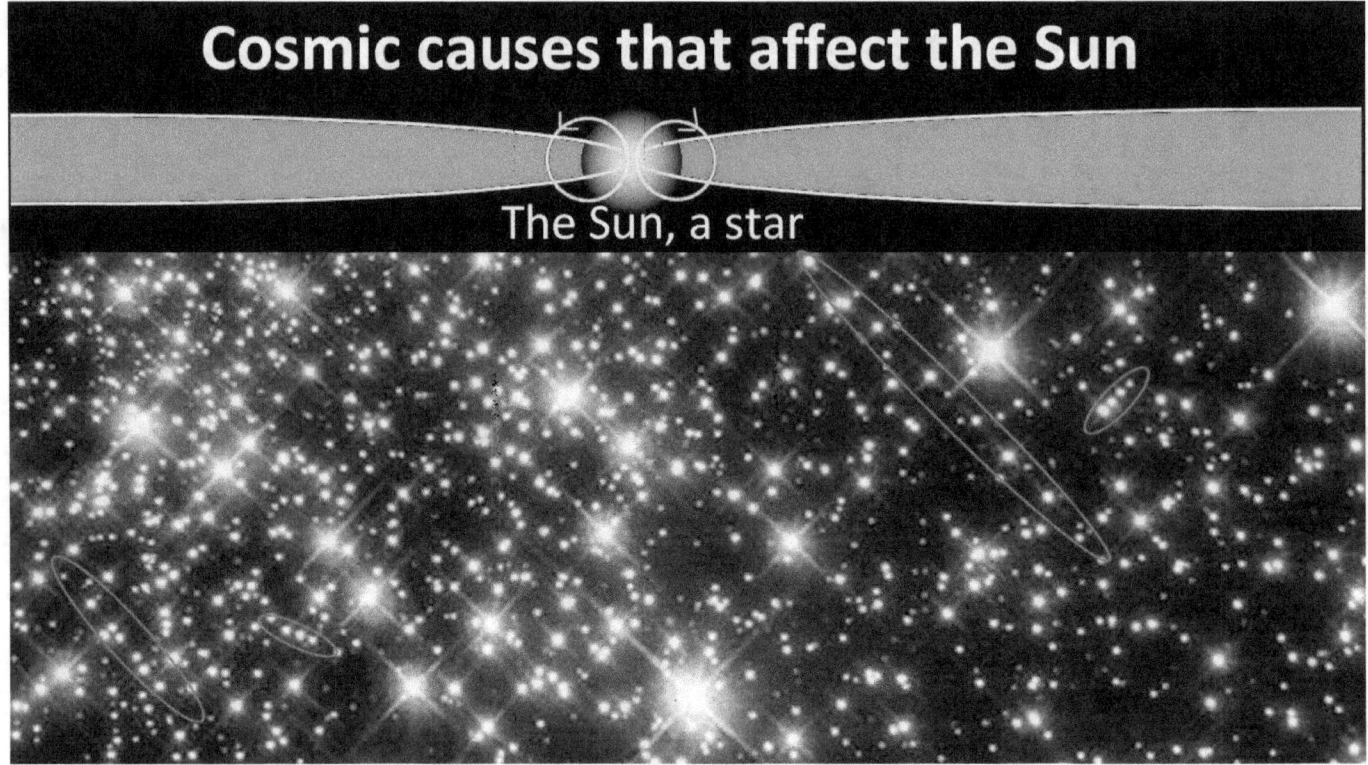

Visual evidence indicates that plasma stars are often lined up into long strings of stars in cosmic space, each of which thereby becomes a node point in the long stream. The plasma that exists in the interstellar background, feeds into the active plasma stream. The background plasma is a feature of the dynamics of the galaxy.

When the plasma flow becomes too weak

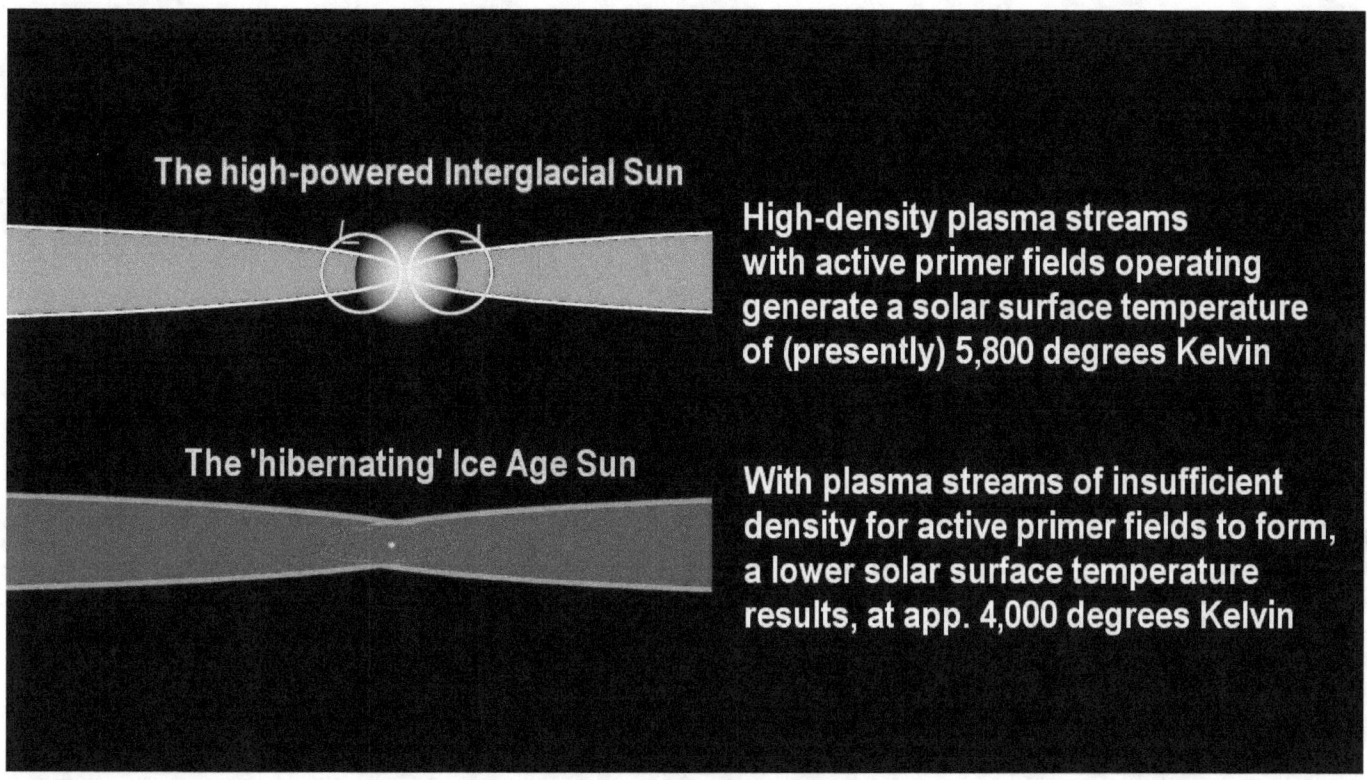

The critical aspect for us, about our Sun as a plasma star being located at a node point of an interstellar stream, is that the primer fields form by themselves by the electromagnetic effect of moving plasma, whenever the conditions exist for this to happen.

In order for the primer fields to form, and to be maintained, the plasma stream must have a minimal rate of flow and density. When the plasma flow becomes too weak for the primer fields to be maintained, the primer fields vanish and their effects vanish with them. In this case the plasma sun looses the concentrated plasma surrounding it, and falls back into a low-power hibernation state that corresponds to the less focused plasma environment that exists without primer fields in operation.

The plasma density of the interstellar stream for our solar system is presently, rapidly diminishing. The threshold will likely be crossed in the 2050s past which the primer fields can no longer be maintained and the Sun drops into hibernation. That's the 'promise' of our Sun as a plasma star.

The deep Ice Age resumes that has not been experienced

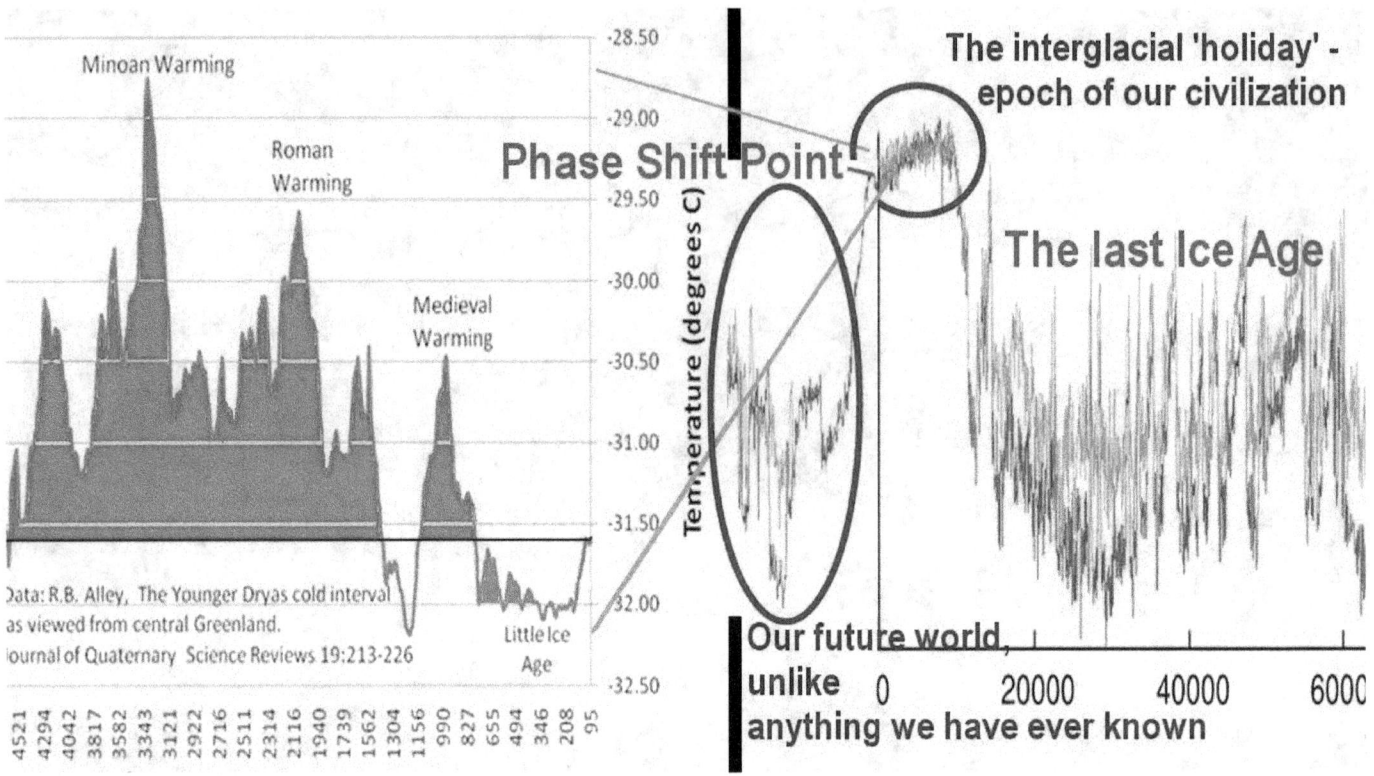

With it the current interglacial period ends, the next glaciation periods begins, the deep Ice Age resumes that has not been experienced in the entire 12,000-year epoch in which our civilization developed.

The hibernating plasma Sun has measured evidence

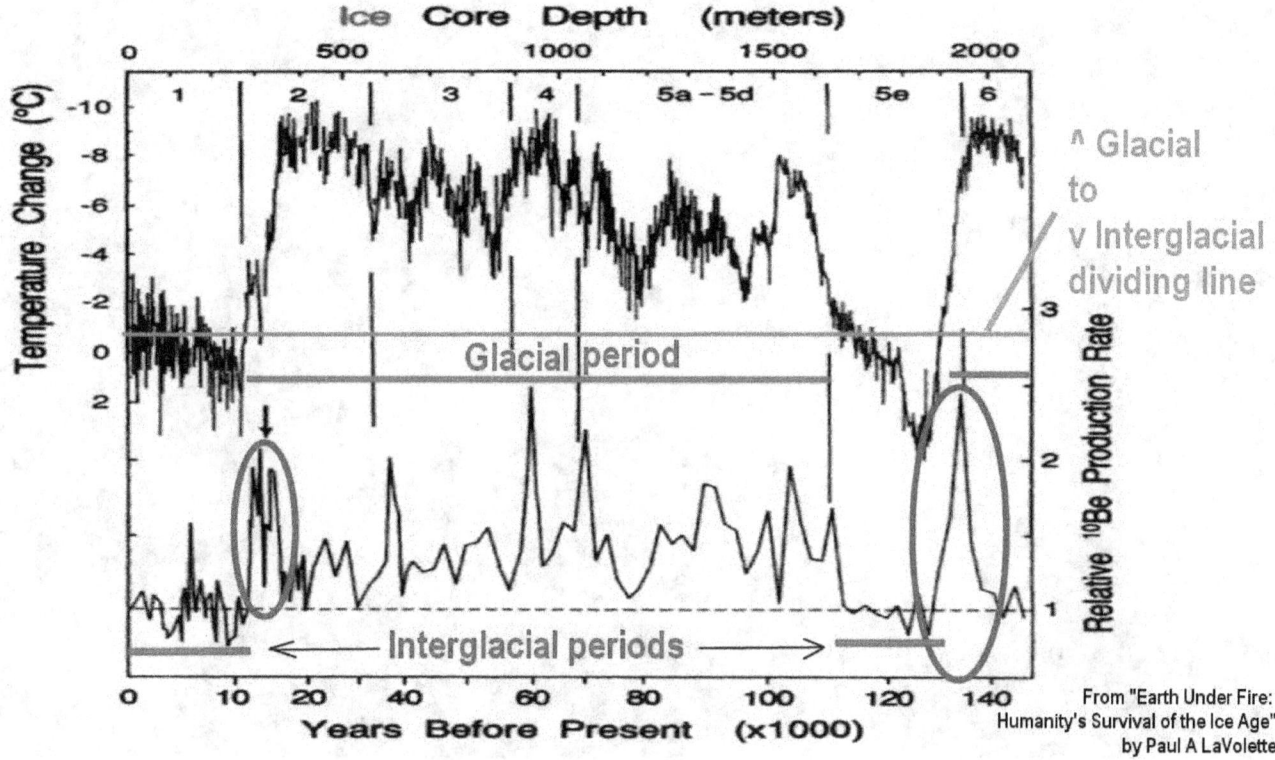

That the hibernating plasma Sun is not a mere theory, but has measured evidence standing behind it, comes to light in ice core data going back in time 150,000 years.

The hibernating Sun, without a plasma mantle around it, emits significantly larger volumes of cosmic-ray flux, which becomes reflected on Earth in high rates of Berillium-10 production in the atmosphere, preserved in ice. The high rate begins and ends with the beginning and ending of the last glacial period.

This proof, delivered by the Sun, of its state as a hibernating plasma star, proves that Ice Ages are caused by the Sun and by what affects it. This proof invalidates all mechanistic theories for the cause of the Ice Ages, because nothing other than a hibernating sun, being a plasma star, can generate the high volume of cosmic-ray flux that is reflected in high Berillium-10 ratios that have been measured, and have been measured precisely aligned with the glaciation period.

The Sun is presently in its interglacial high-powered mode

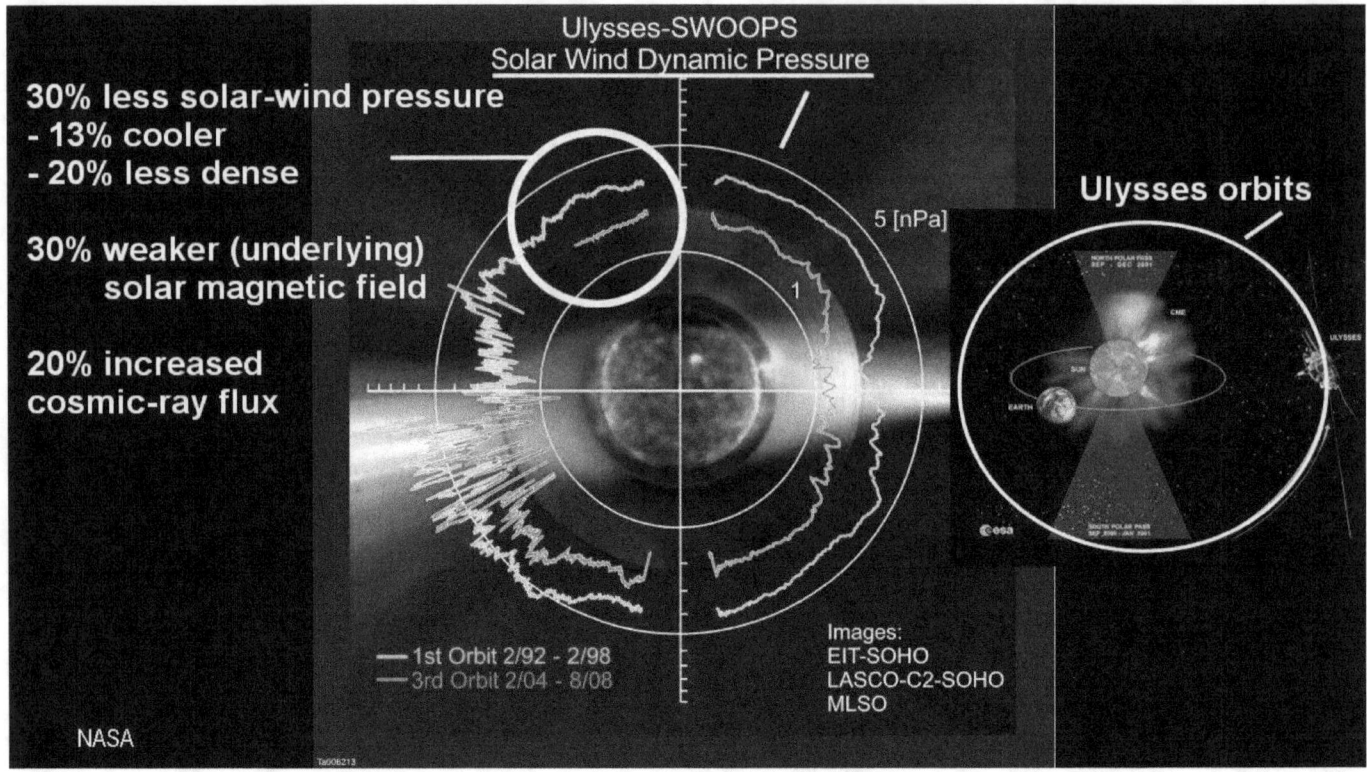

That the Sun is presently in its interglacial high-powered mode, is evident in the measurements produced by the NASA and ESO spacecraft, Ulysses, as it orbited over the polar regions of the Sun.

One of the objectives of the space mission was to measure the distribution of solar-wind pressure around the Sun in all directions, outside the ecliptic.

A complete void of solar-wind movement over the poles

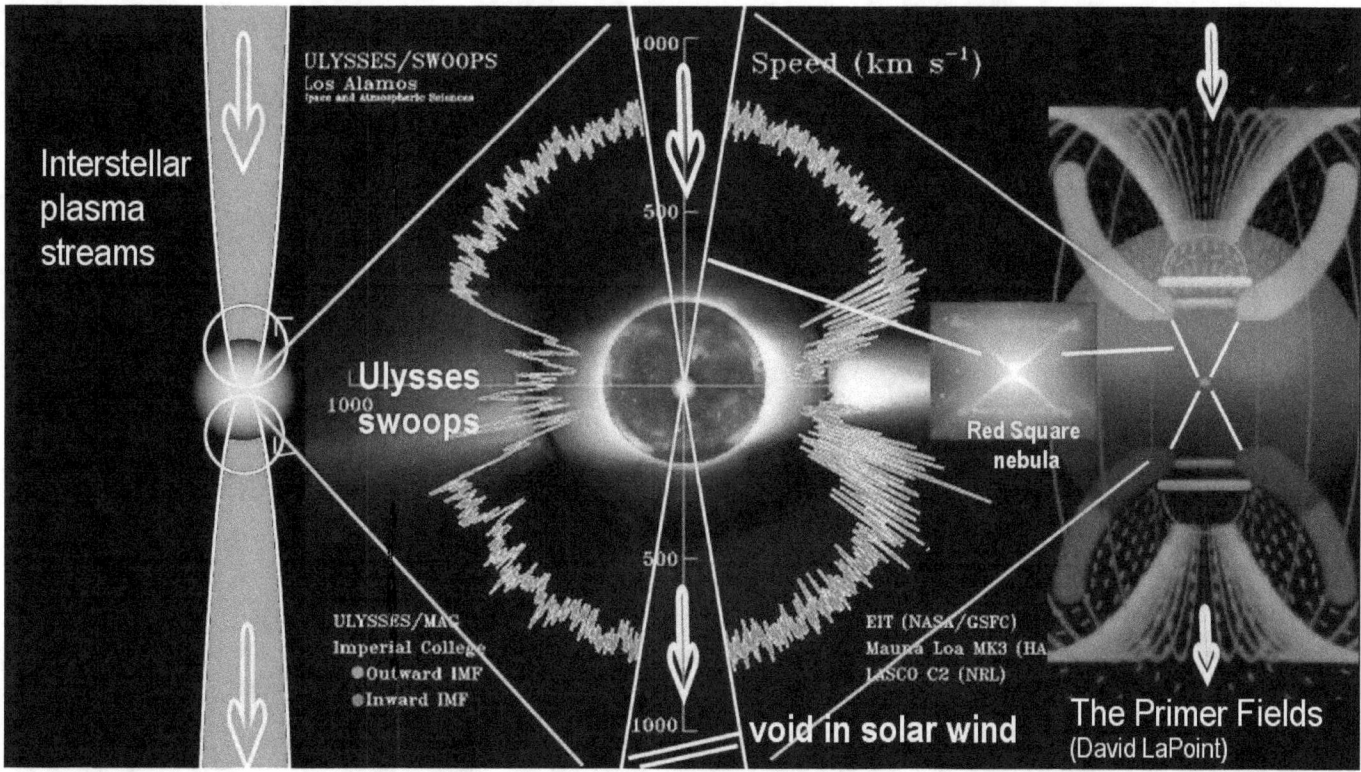

Here it gets interesting. The satellite encountered a complete void in its measurements, of the solar-wind movement over the poles, in the exact area where the plasma inflow from the primer fields would be logically located. The Ulysses measurements add another item of proof, thereby, that our Sun is a plasma star, and all stars likewise. It also adds to the recognition that the galaxy as a whole, in which we are located, is a plasma-powered structure.

Two gigantic plasma structures

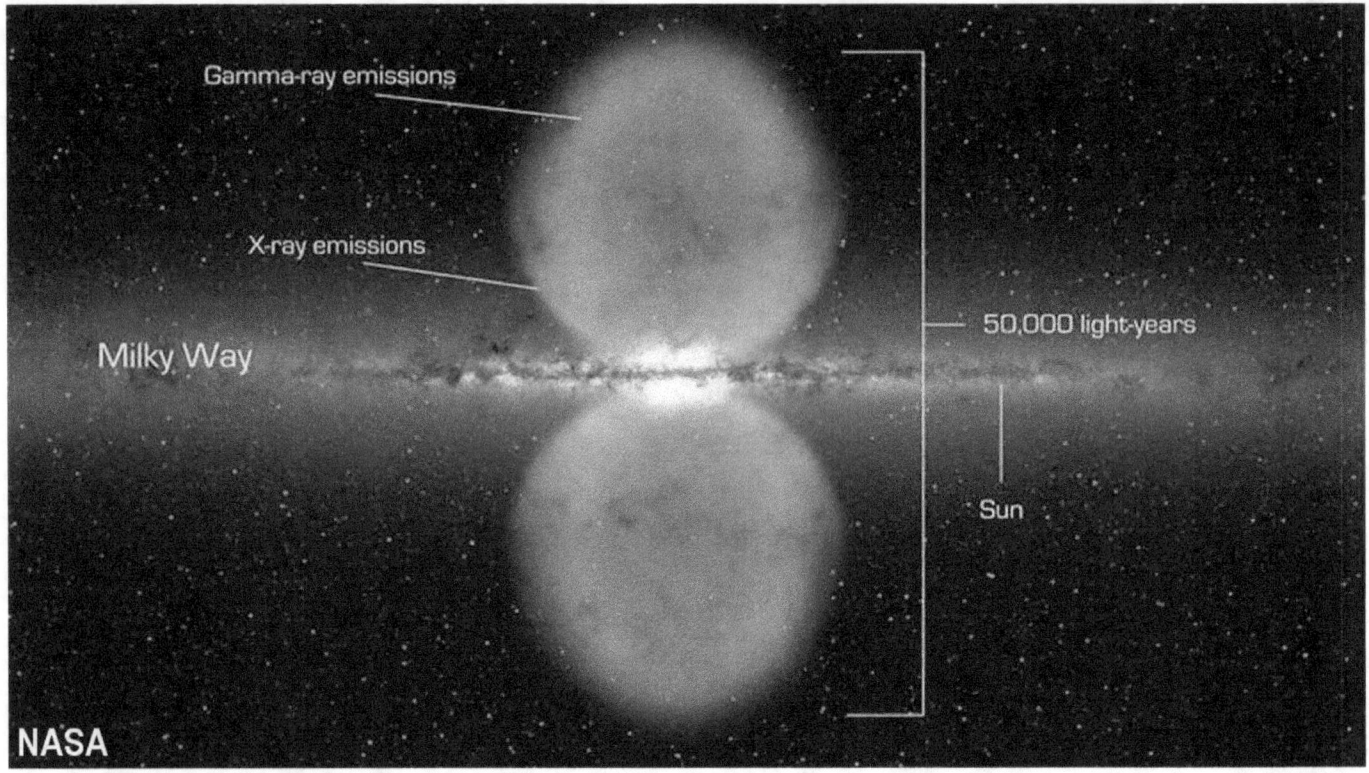

NASA has discovered the existence of two gigantic plasma structures above and below the galactic disk.

The plasma structures are naturally visible in gamma-ray and x-ray light

The structures are reminiscent to a plasma feature observed in high-energy plasma discharge experiment. In the lab the plasma geometry was made artificially visible. The plasma structures above and below the galaxy are naturally visible in gamma-ray and x-ray light.

The similarity in shape and location

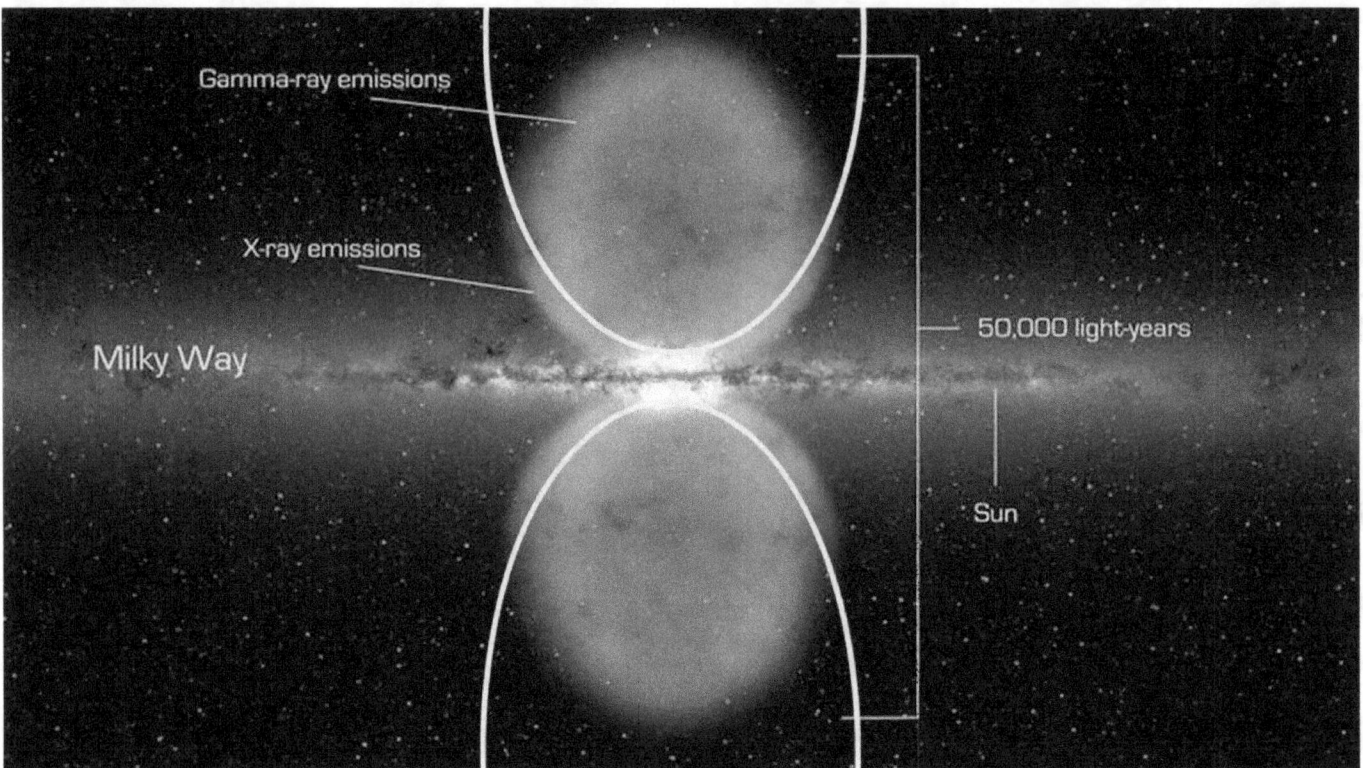

The similarity in shape and location renders the galactic plasma structures as confinement domes of galactic-scale primer fields, and the galaxy as a node point in intergalactic plasma streams.

The entire galaxy as a part of a node-point structure

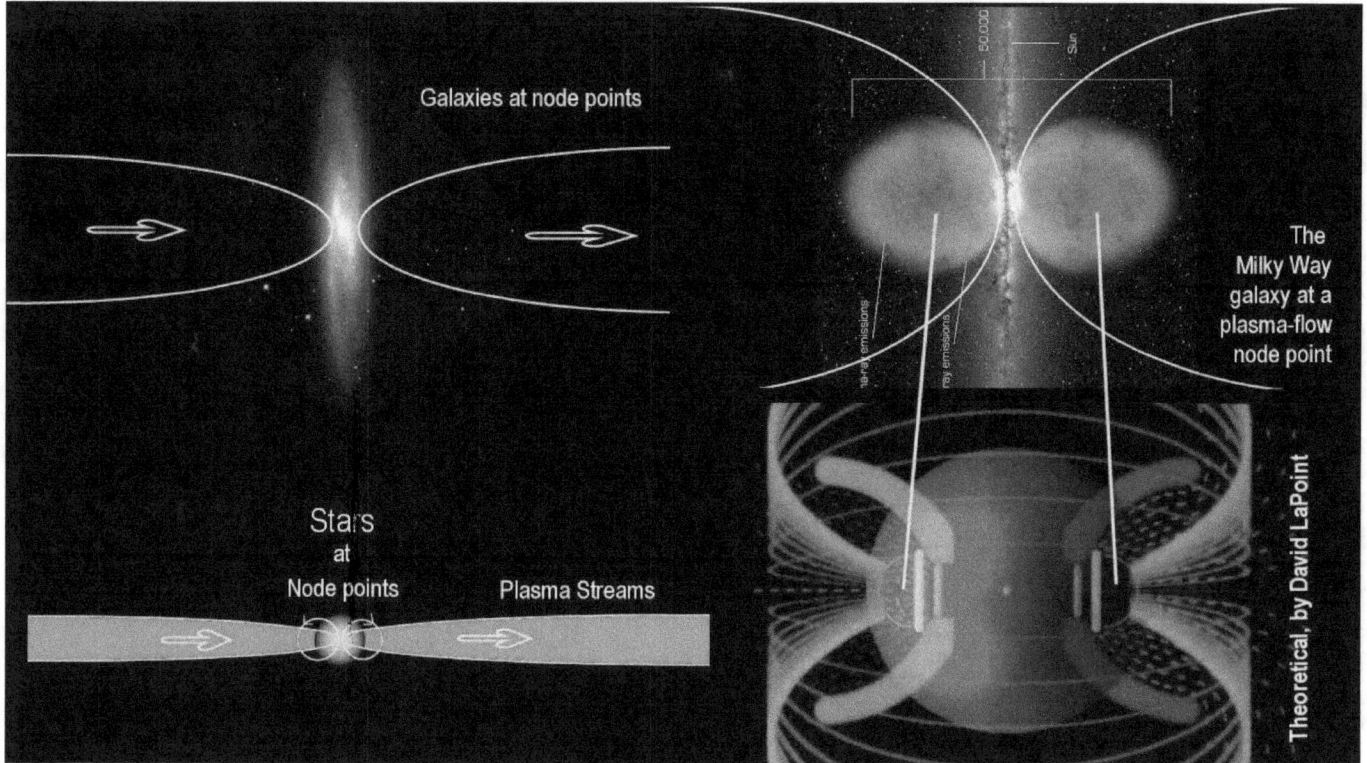

The discovery renders the entire galaxy as a part of a node-point structure between immensely long intergalactic plasma streams, just as our solar system and its primer fields is a node point structure between interstellar plasma streams.

The basic operational principle is the same in both cases, expressed on vastly different scales.

The discovery thereby obsoletes the Big Bang theory

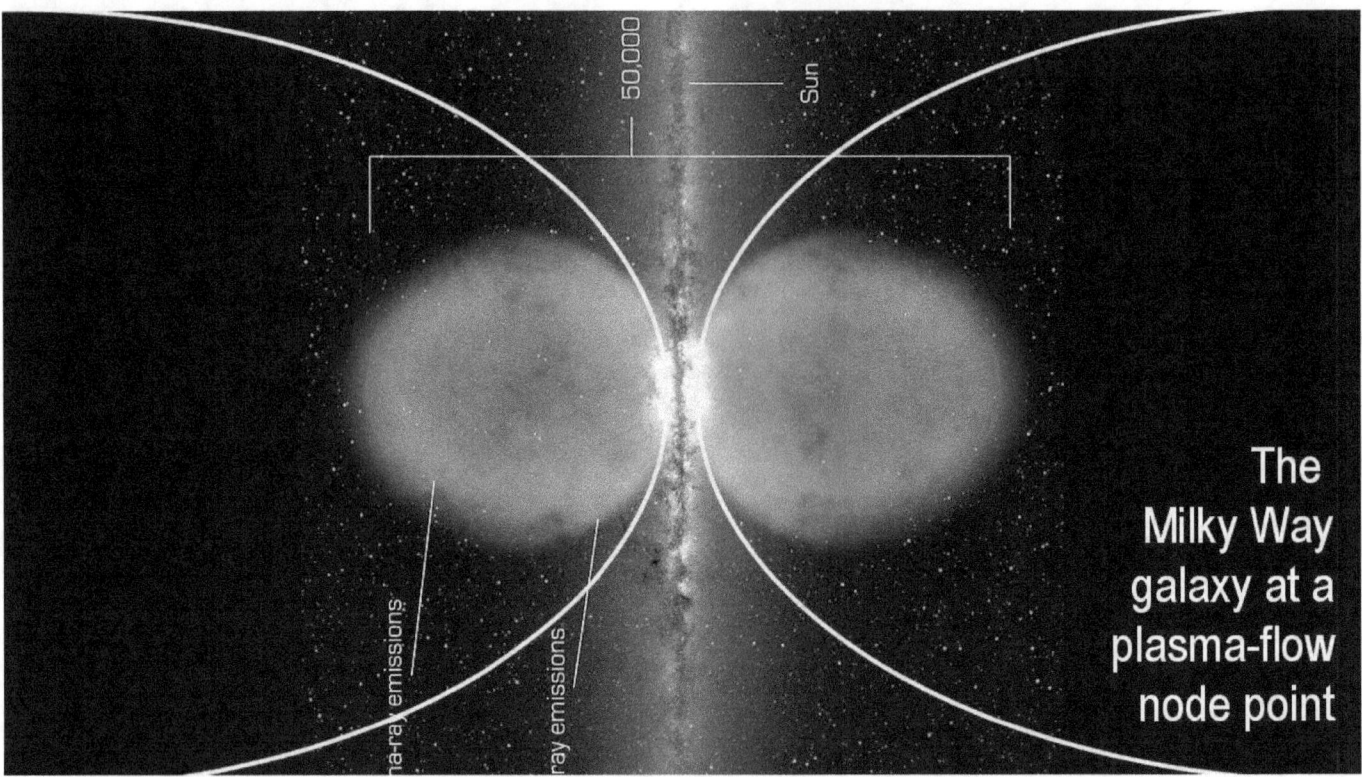

The discovery of the galactic plasma structures renders our entire galaxy a sea of 400 billion plasma-powered, plasma stars, and likewise so, all galaxies.

The discovery thereby obsoletes the Big Bang theory, which had been invented to theorize the origin for the hydrogen gas for the gas sun model. This has been made obsolete by NASA's galactic-plasma discovery - though many like to dispute this, which is plainly obvious.

The origin of the cosmic plasma streams

The origin of the cosmic plasma streams that power the plasma galaxies, which all galaxies are, is actually much more elegant and rationally acceptable than the Big Bang theory that regards the universe to be the result of a cosmic explosion.

The theoretical physicist David Bohm

The theoretical physicist David Bohm, whom Albert Einstein is said to have referred to as his successor, suggests that nothing is actually created.

Empty space as not at all empty

David Bohm sees apparently empty space as not at all empty, but as a sea of immensely dense latent energy that has an implicate order and an explicate order.

The explicate order may be like ripples on the surface

The explicate order may be like ripples on the surface of water. Protons and electrons may be like ripples on the surface of energy.

Electrons and protons constructs of quarks

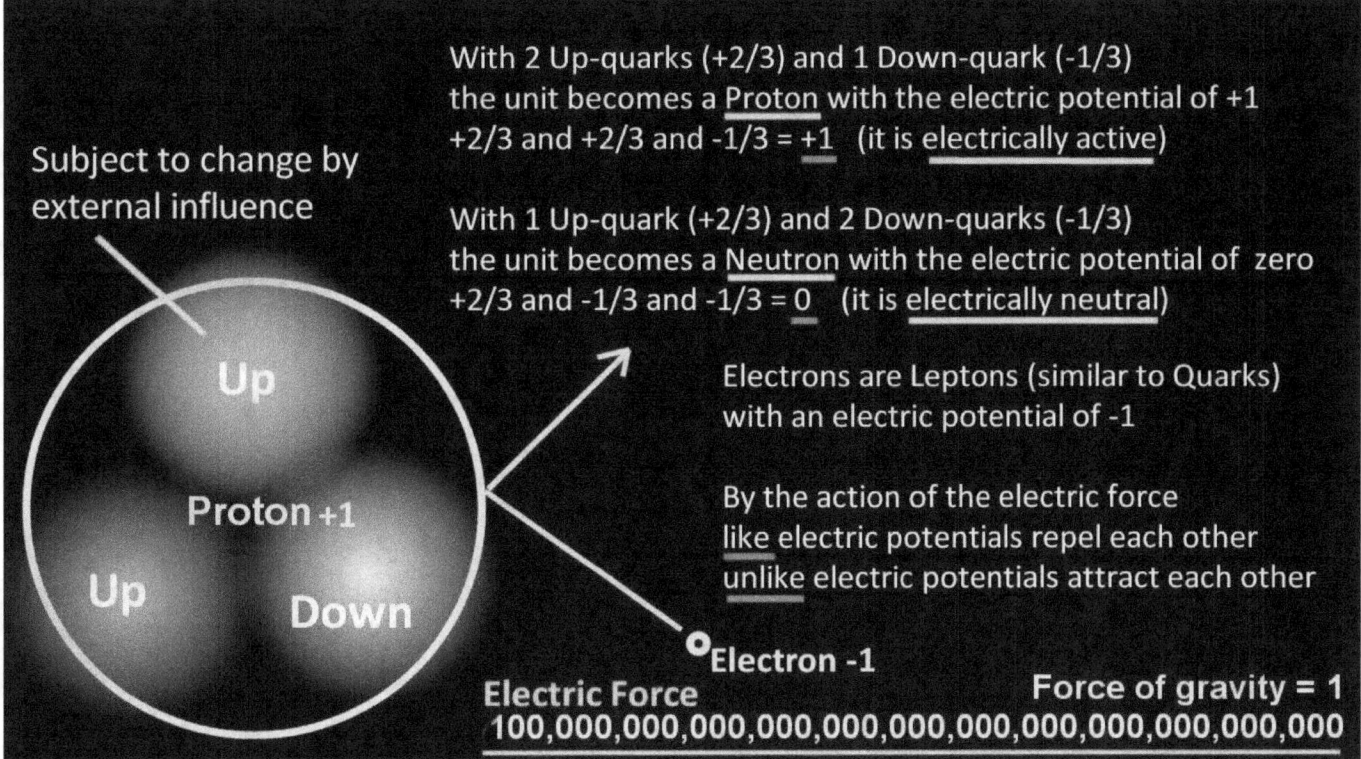

It is well understood in nuclear physics that electrons and protons are not basic particles in themselves, but are theorized to be the constructs of quarks, which are theorized themselves, to be but moving points of energy.

The cycle of plasma in the universe is a gigantic cycle

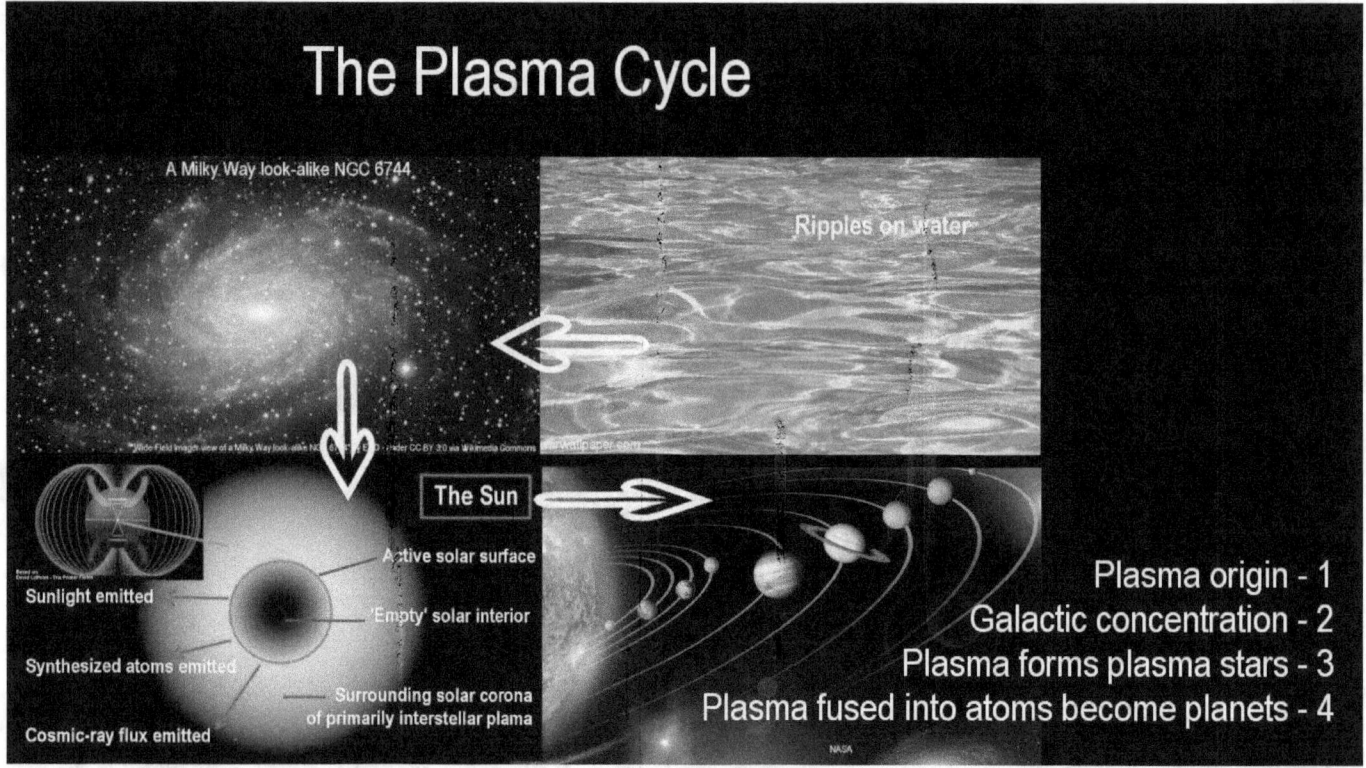

Plasma origin - 1

Galactic concentration - 2

Plasma forms plasma stars - 3

Plasma fused into atoms become planets - 4

The cycle of plasma in the universe is a gigantic cycle. It is an eternal cycle that unfolds in 4 phases simultaneously.

In the first phase, the origin of the plasma that the visible universe is constructed of, unfolds. It unfolds as an explicate phenomenon of ripples on latent energy that fills all space.

In the second phase, the unfolding cosmic plasma background, nourishes intergalactic plasma streams and the galaxies at their node points, whereby the cosmic plasma becomes densely concentrated.

In the third phase, the concentrated plasma that is flowing in the galactic system creates and powers the galaxy's stars, which are plasma stars.

In the fourth phase, the plasma stars fuse plasma into electrically neutral atoms. The process provides the sink effect that keeps the plasma streams in the galaxy, and thereby in the universe, flowing. The process of atomic synthesis also provides the atomic mass for all the planets, and generates the cosmic ray flux and the light and heat that nourish life.

Does all this sound too exotic? I would say, it doesn't.

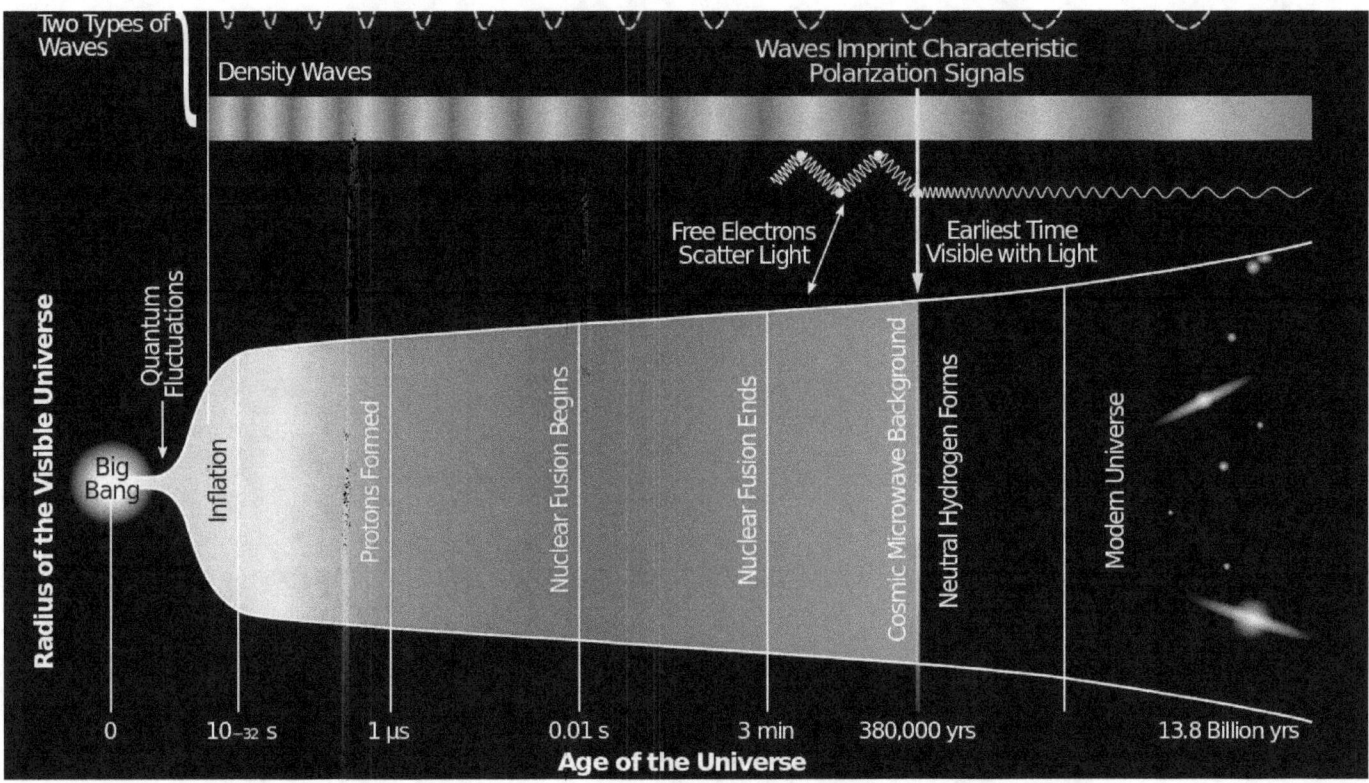

Does all this sound too exotic? I would say, it doesn't. It is far less exotic than the Big Bang explosion myth that is supposed to have created the universe?

Actually, the plasma cycle is not exotic at all

Actually, the plasma cycle is not exotic at all, but is instead extremely relevant to our living on Earth, in a highly practical manner.

The greatest existential challenge

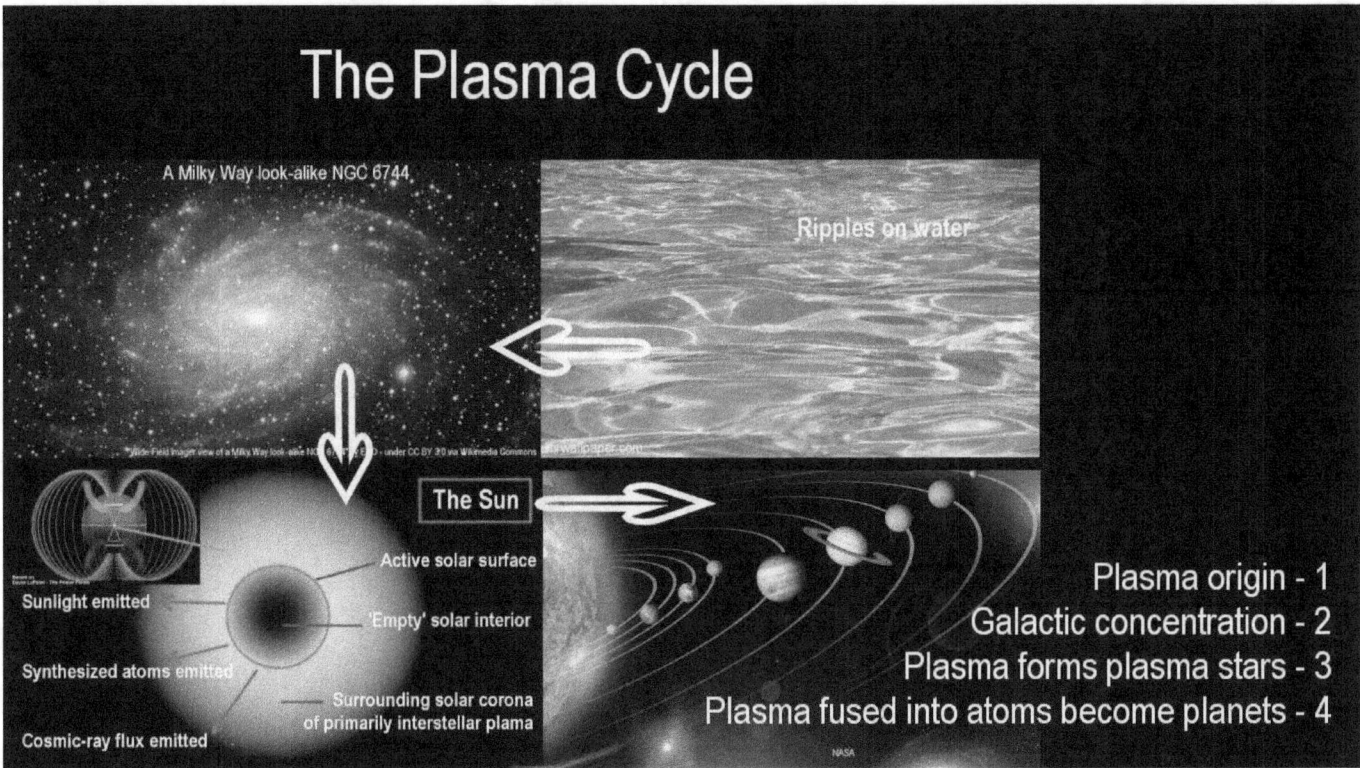

This is so, because the plasma universe, the plasma galaxy, the plasma Sun, and the Earth under the Sun, stand at the center of the greatest existential challenge in our world, in possibly the entire history of humanity.

This challenge, which is immense in scope, unfolds by the galactic plasma system, and the solar plasma system being interlinked, and by them being linked with our living by their effect on the climate on Earth.

Resonance effects of the intergalactic plasma streams

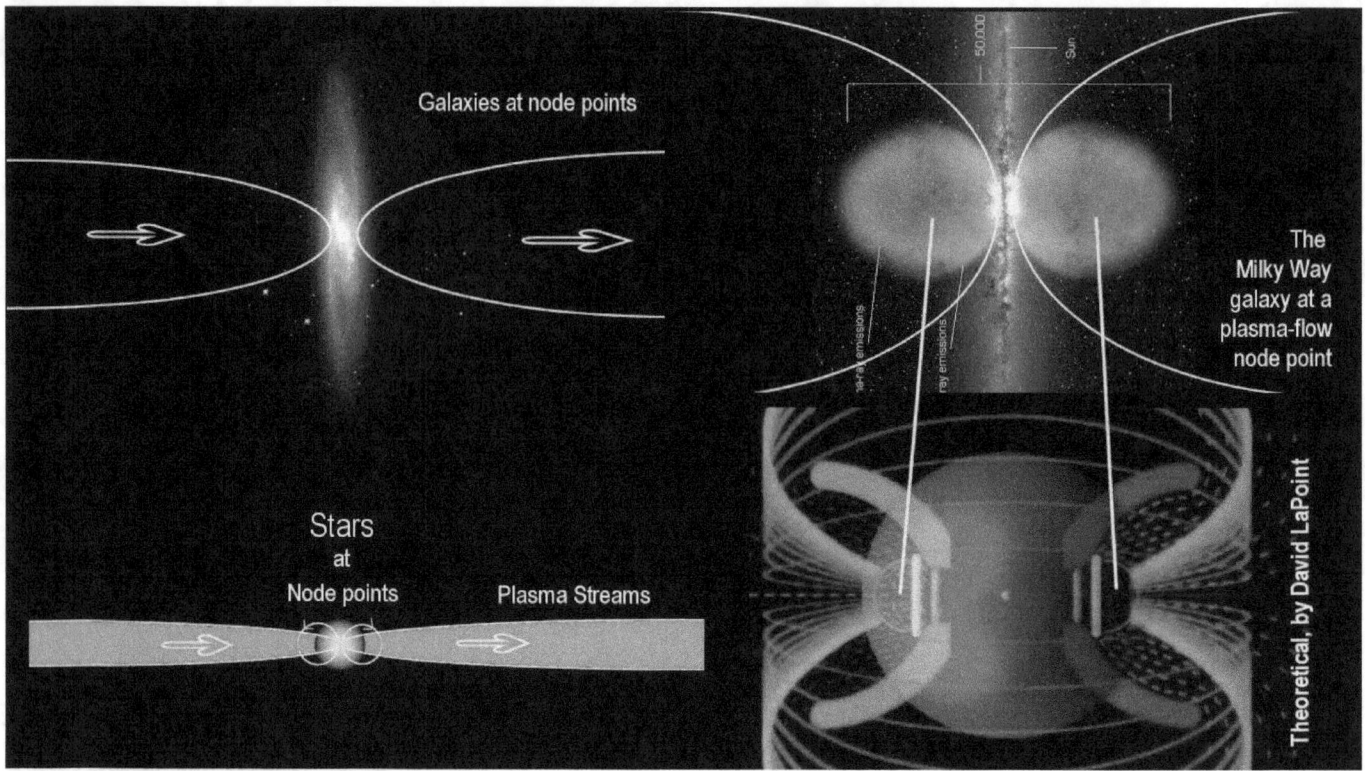

This is so, because the long-cycle resonance effects of the intergalactic plasma streams, cause fluctuations of the density in the galaxy that directly affect our climate on Earth, because the galaxy provides the background stage on with our Sun's own activity fluctuations unfold, with all their climate-change effects, large and small, that even include the ice ages and the one that is now on the near horizon.

At the coldest level of the last 440 million years

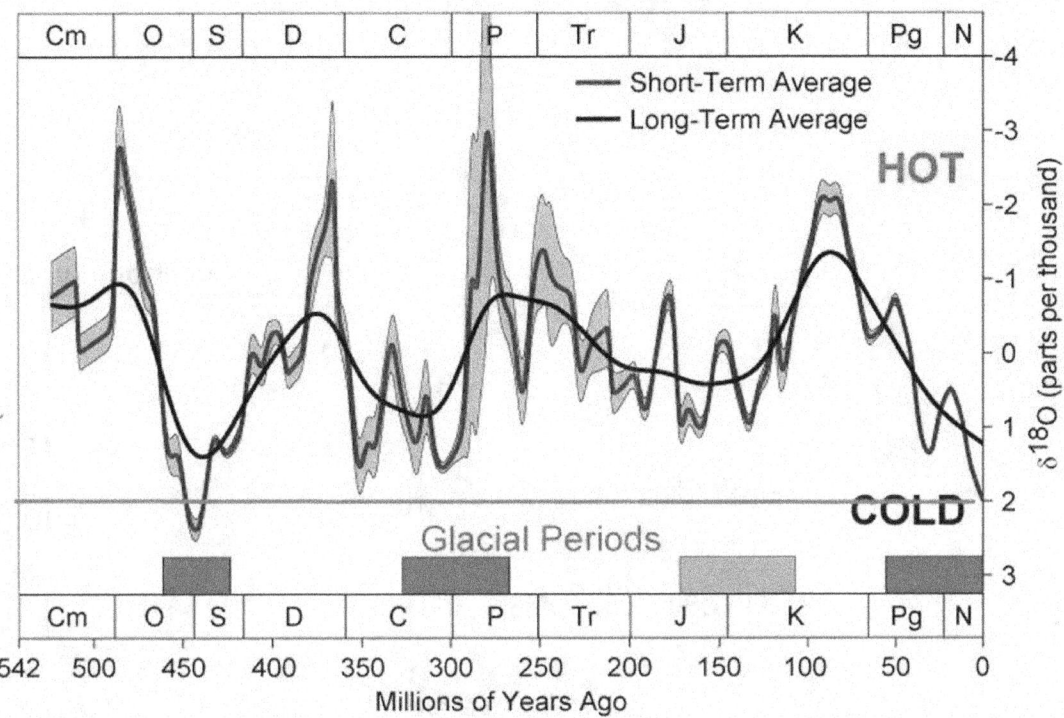

On the phanerozoic time scale our climate is presently at the coldest level of the last 440 million years. We are way colder now than in the time when Antarctica froze up for a second time 12 million years ago, and we are getting colder still.

We are now 2 million years into the ice age cycles epoch

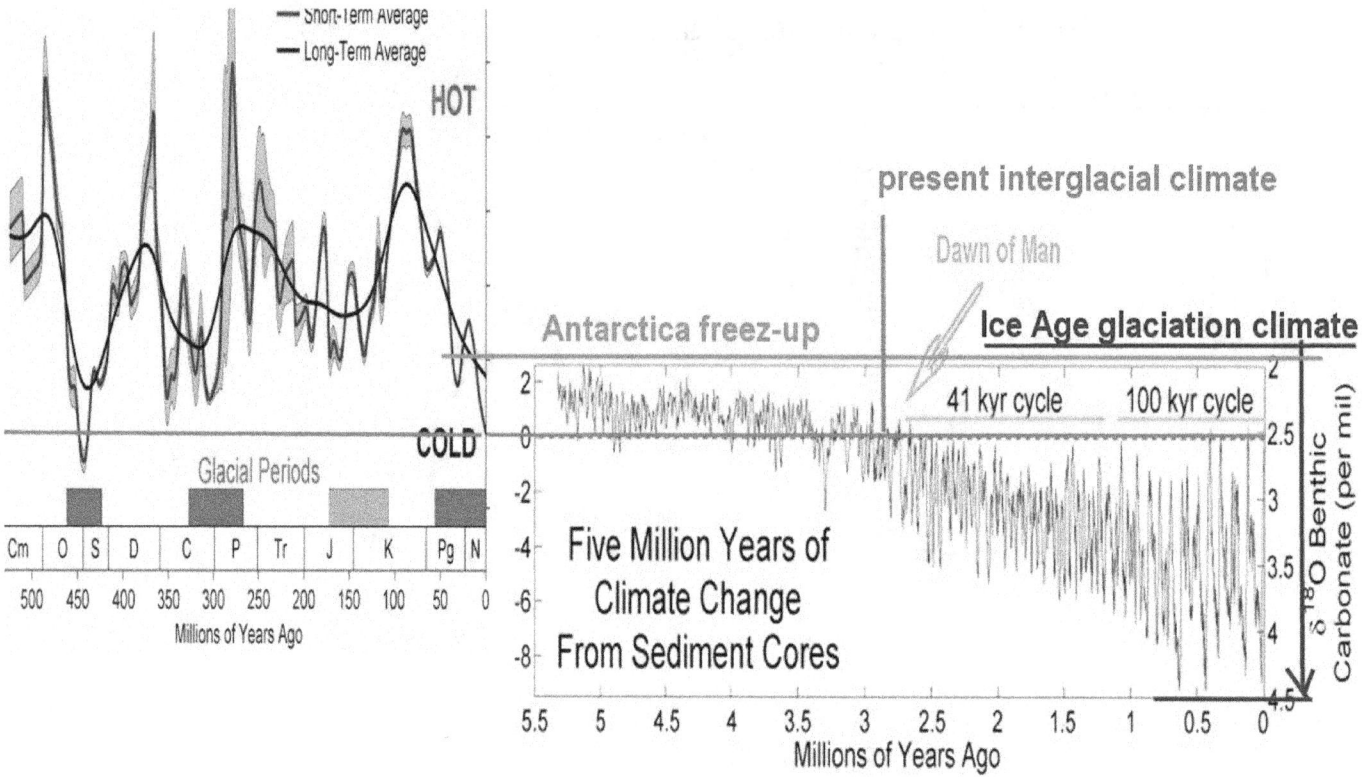

Then, ten million years after Antarctica froze up, the Ice Age glaciation cycles began. We are now 2 million years into the ice age cycles epoch, in which the climate that we presently enjoy, the interglacial climate, is an anomaly that barely lasts for slightly over 10% of the Ice Age cycle.

Plasma streams have a built-in resonance

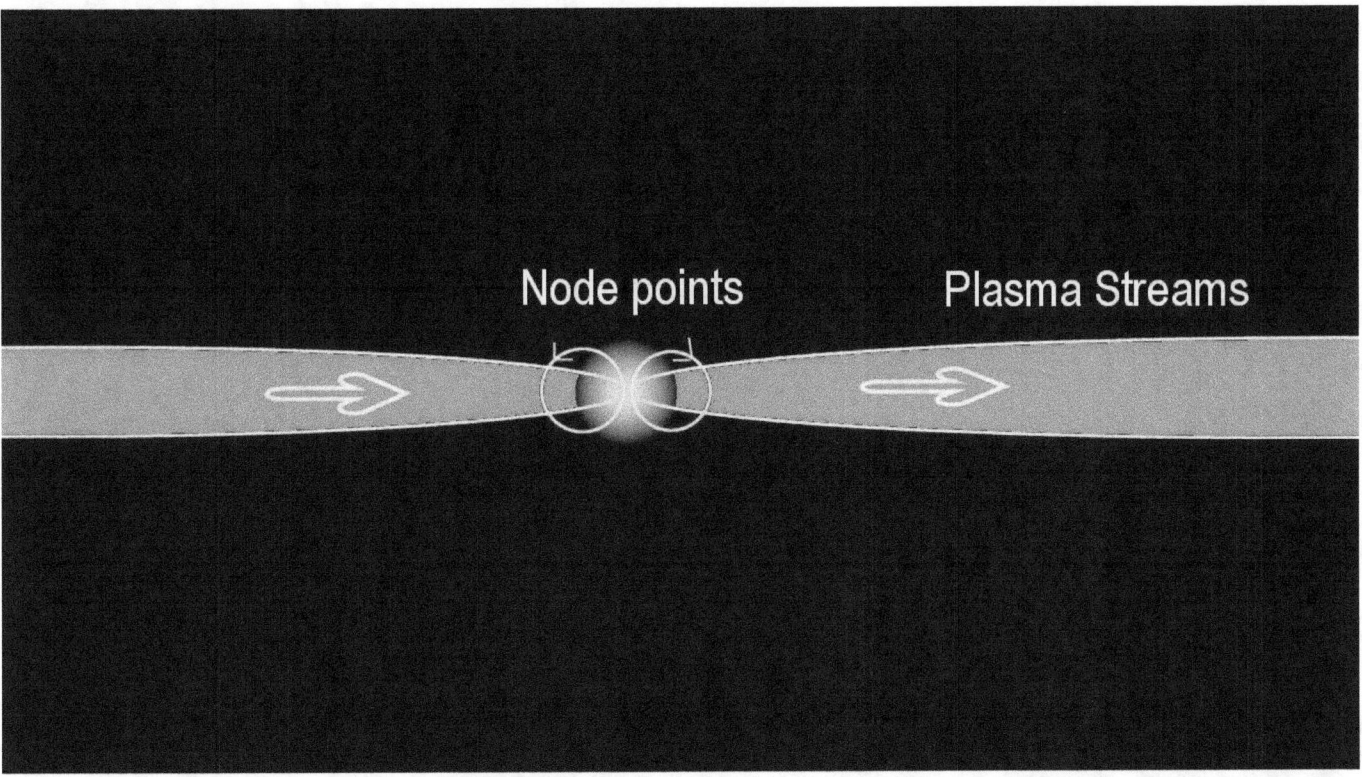

The anomaly is caused by the electromagnetic resonance, this time, in the interstellar plasma streams, just as we find such resonances in intergalactic plasma streams.

Plasma streams are electric in nature, and elastic. They have a built-in resonance according to their length and their plasma density.

Solar cycles are essentially resonance features

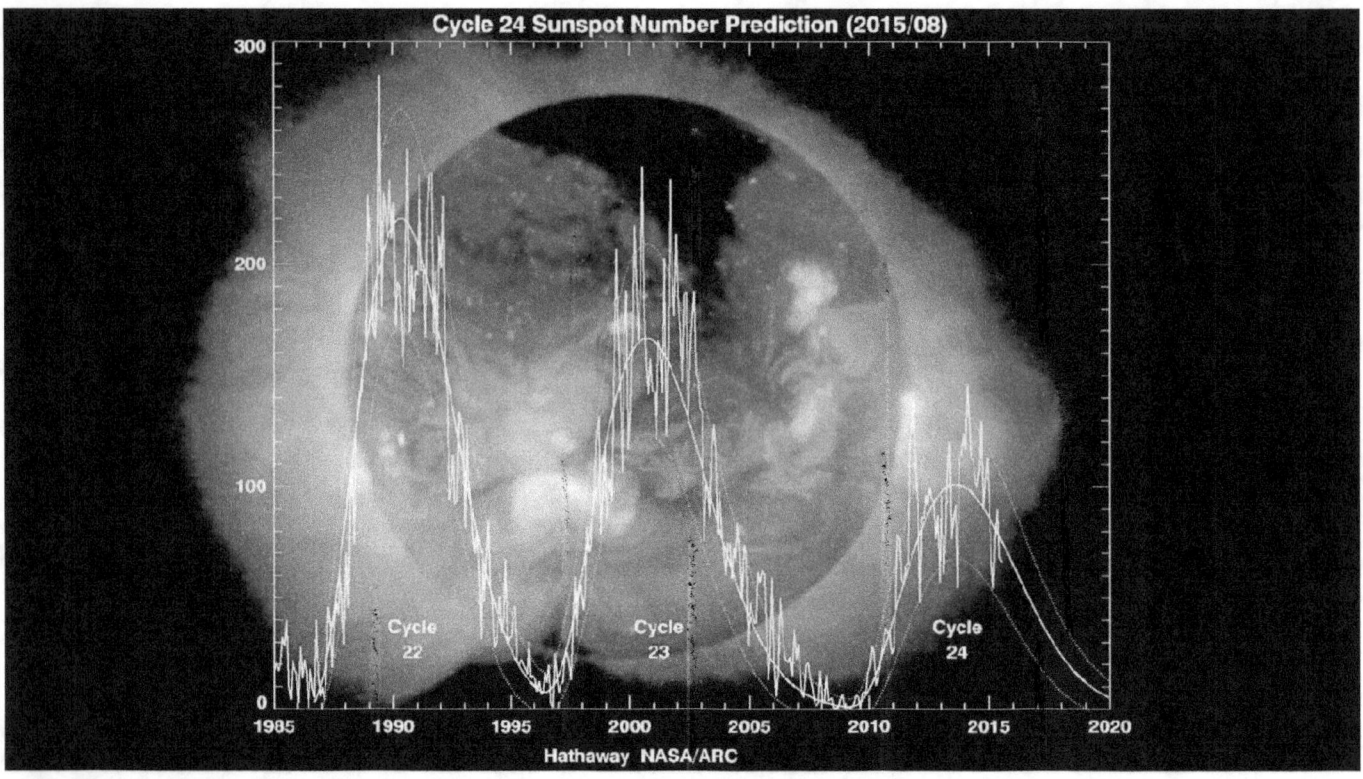

The built-in resonance causes the density of the interstellar plasma streams to fluctuate, similar to the fluctuations of the solar cycles that we see plotted here by sunspot numbers, which are essentially likewise resonance features.

The collapse of the primer fields happens at a lower density

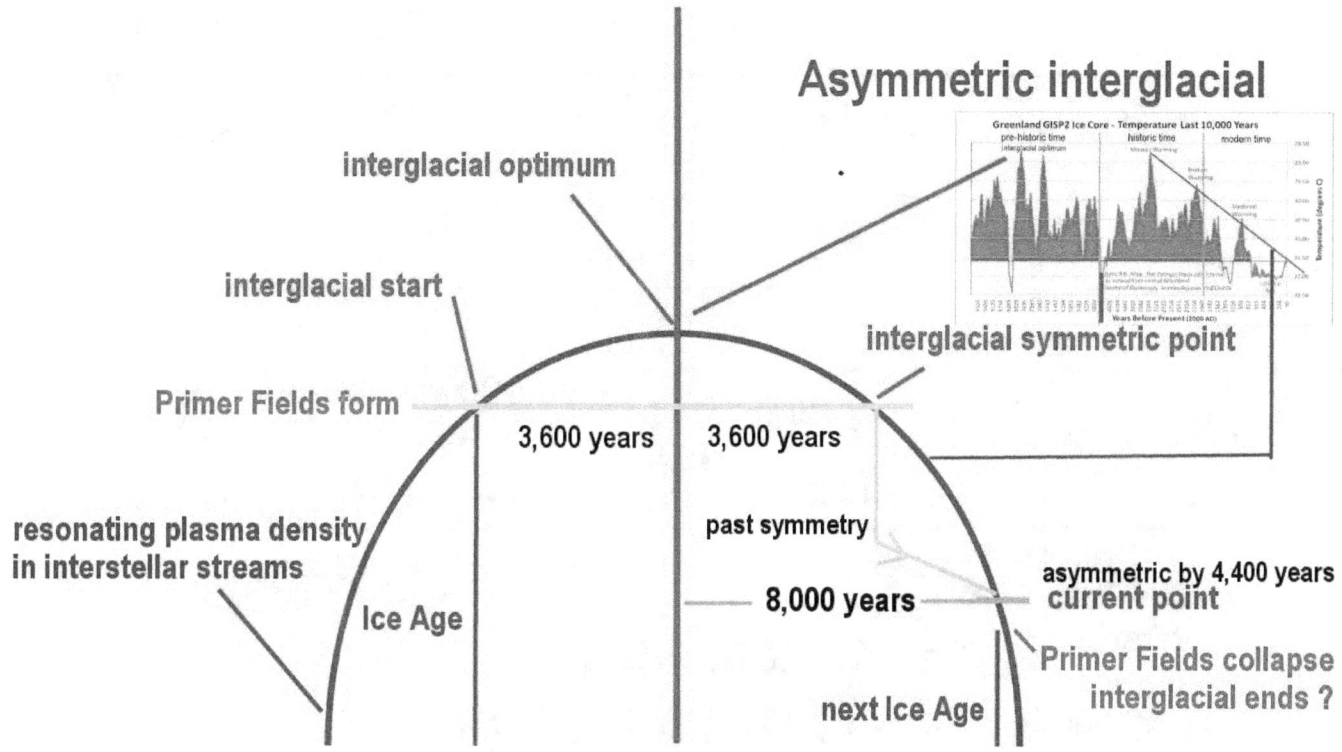

Interglacial periods result when the density in the interstellar plasma stream for our Sun meets the requirement for primer fields to form, which, when active, focus concentrated plasma onto our plasma Sun and enable it to operate in high-powered mode.

The high-powered mode, that is the interglacial mode, will last until the primer fields collapse again. The collapse of the primer fields, however, happens at a lower density level than the turn-on level.

High-powered mode for roughly 10% of the Ice Age Cycle

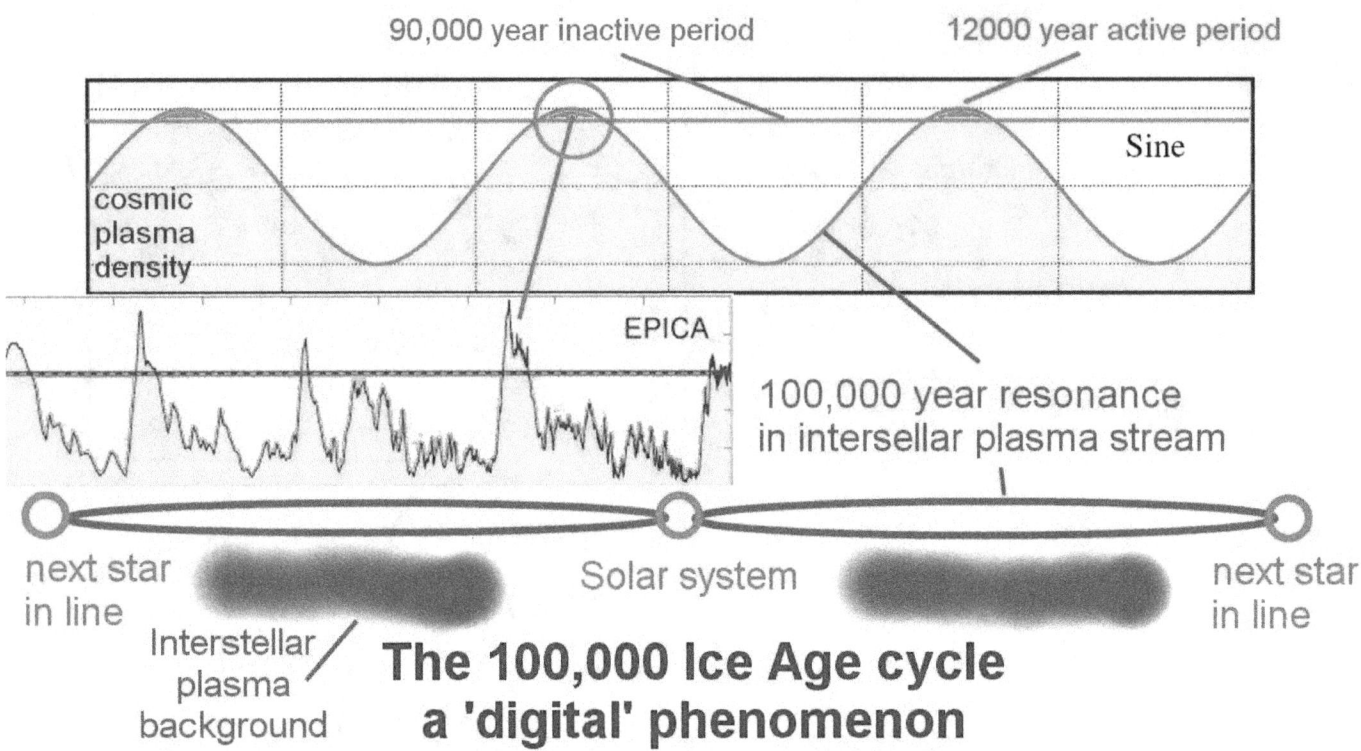

By these dynamics the plasma Sun will be in its interglacial high-powered mode for roughly 10% of the Ice Age Cycle, according to modern measurements.

The interglacial period is asymmetric in nature

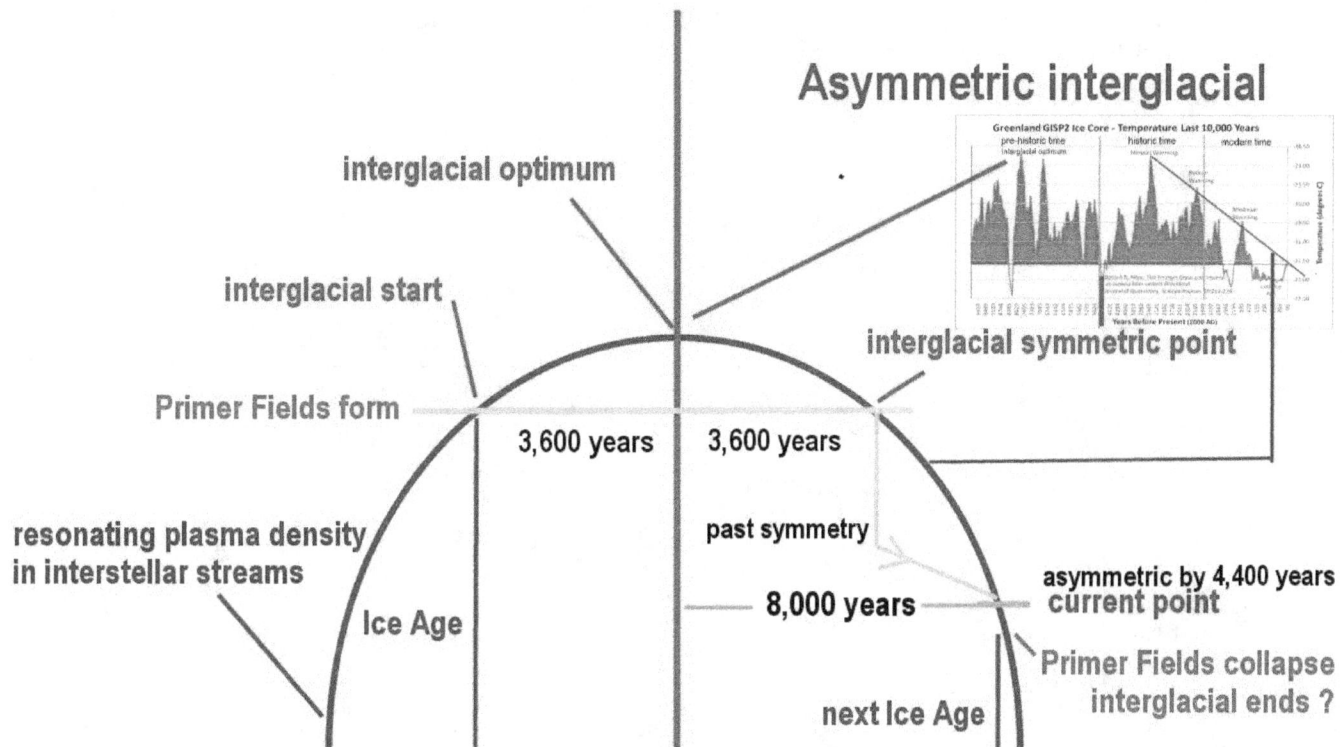

Because the forming and collapse of the primer fields occur at different density levels, the interglacial period is asymmetric in nature, which puts the collapse point onto a steeper slope of the resonance pulse. This means that the interglacial climate is diminishing at a more rapid rate now, even collapsing in some respects.

The interglacial is accelerating towards its end

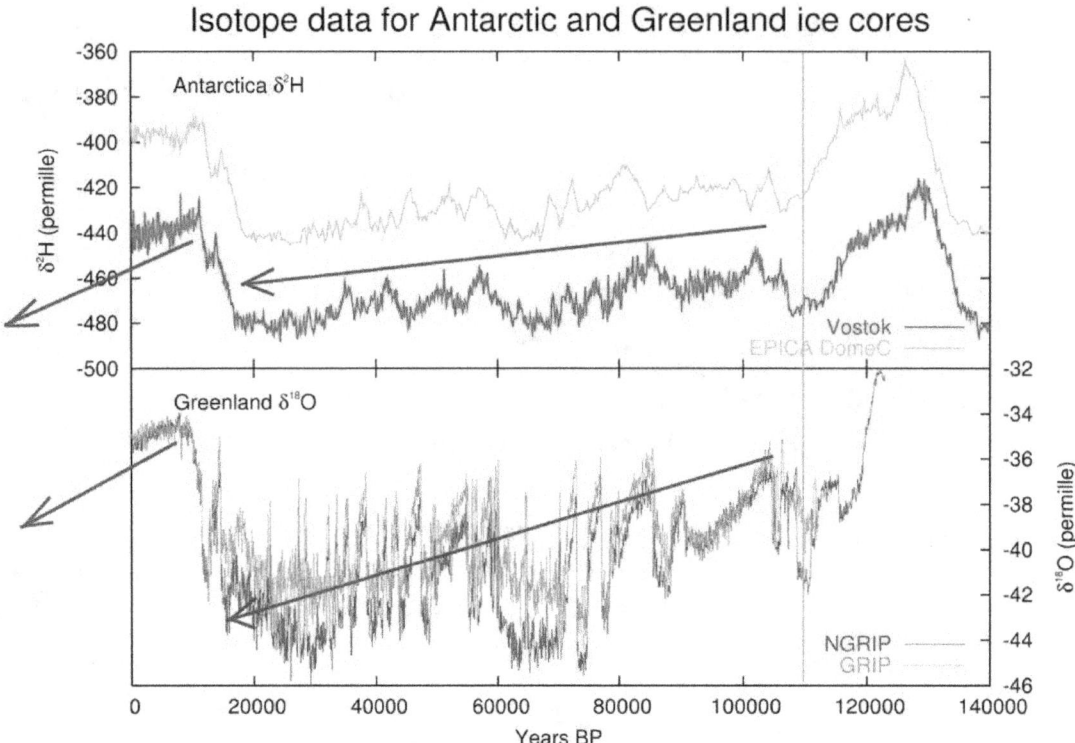

At the present stage, our brief anomaly of wonderfully warm climates, which has been diminishing since the interglacial optimum 8,000 years ago, is now accelerating by the effects of its asymmetry. It is accelerating towards its end.

Alarm bells should be ringing.

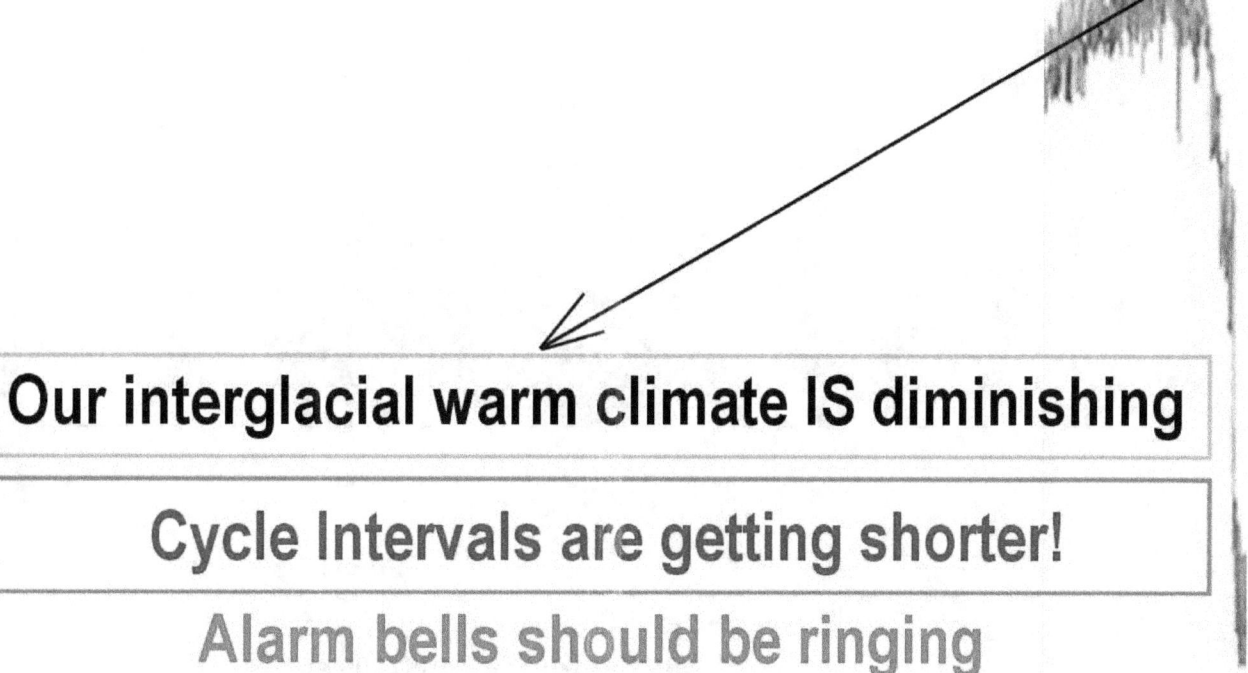

That's what we see reflected in the ice core measurements. We are nearing the end of a wonderful, but brief, anomaly. The interstellar plasma streams that power are climate are getting so dramatically weaker that related effects are popping up everywhere. Alarm bells should be ringing.

The first major effect that happened

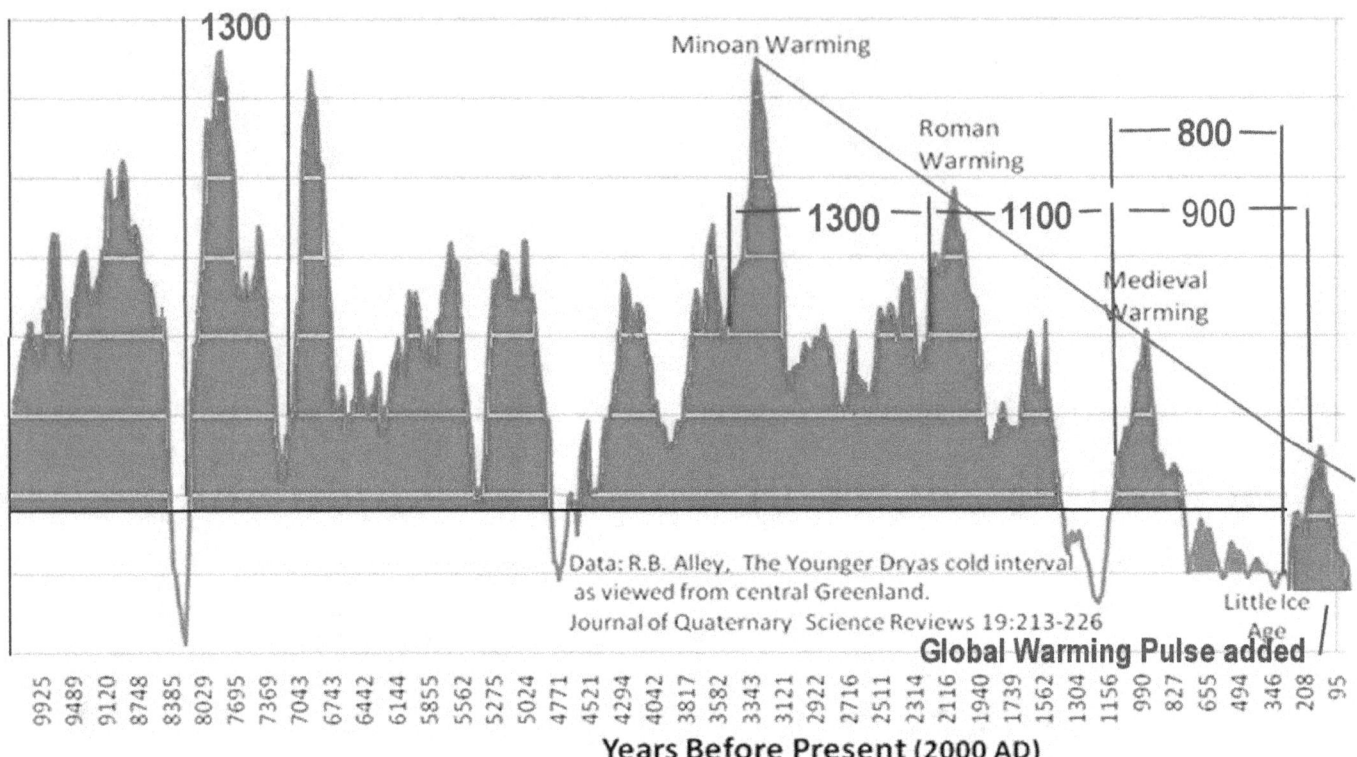

The first major effect that happened, was that the big warming cycles and their intervals began to diminish.

We saw the shorter cycles diminishing likewise

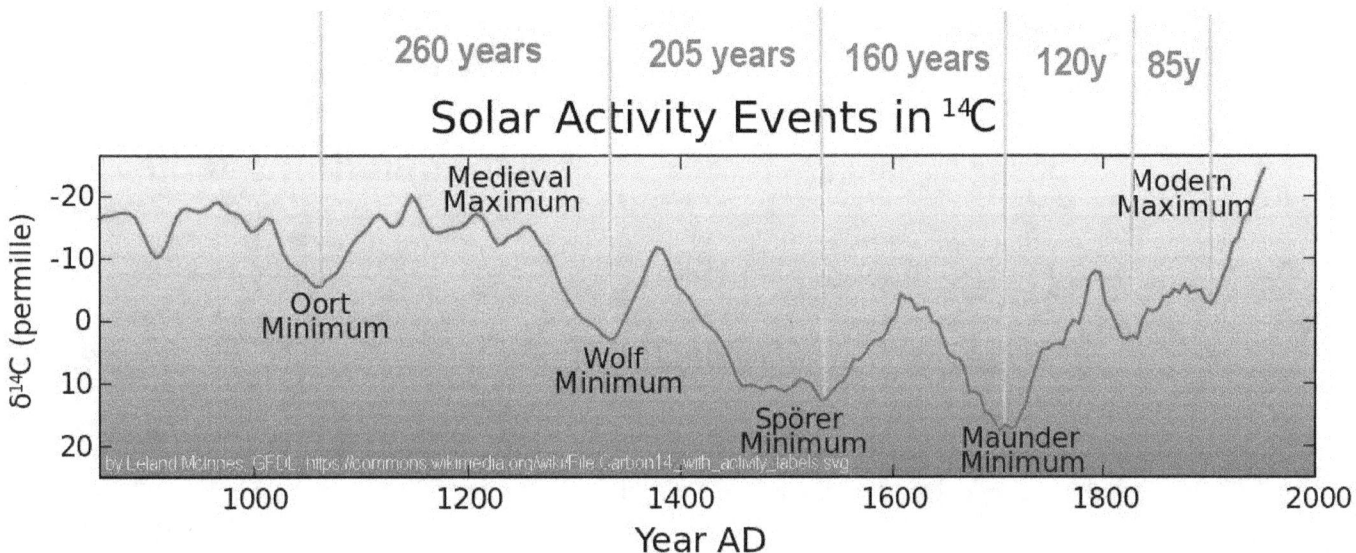

Thereafter, we saw the shorter cycles that the Little Ice Age cycles were a part of, diminishing likewise both in amplitude and in the length of their intervals.

Intervals getting shorter at a rate of geometric progression

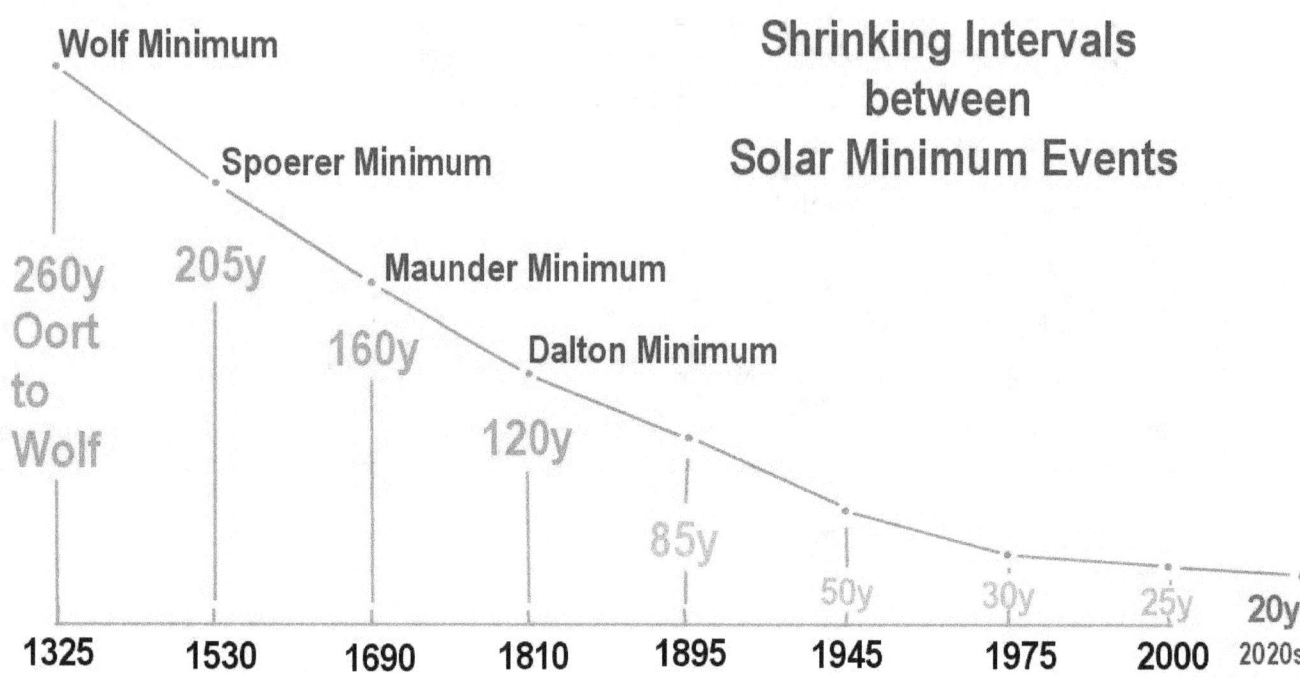

We saw their intervals getting shorter at a rate of geometric progression. There is nothing much left of them now as we are fast approaching the end of the line while the Earth keeps on getting colder. That's the face of the weakening plasma Sun.

Many of the dynamic features are presently loosing ground

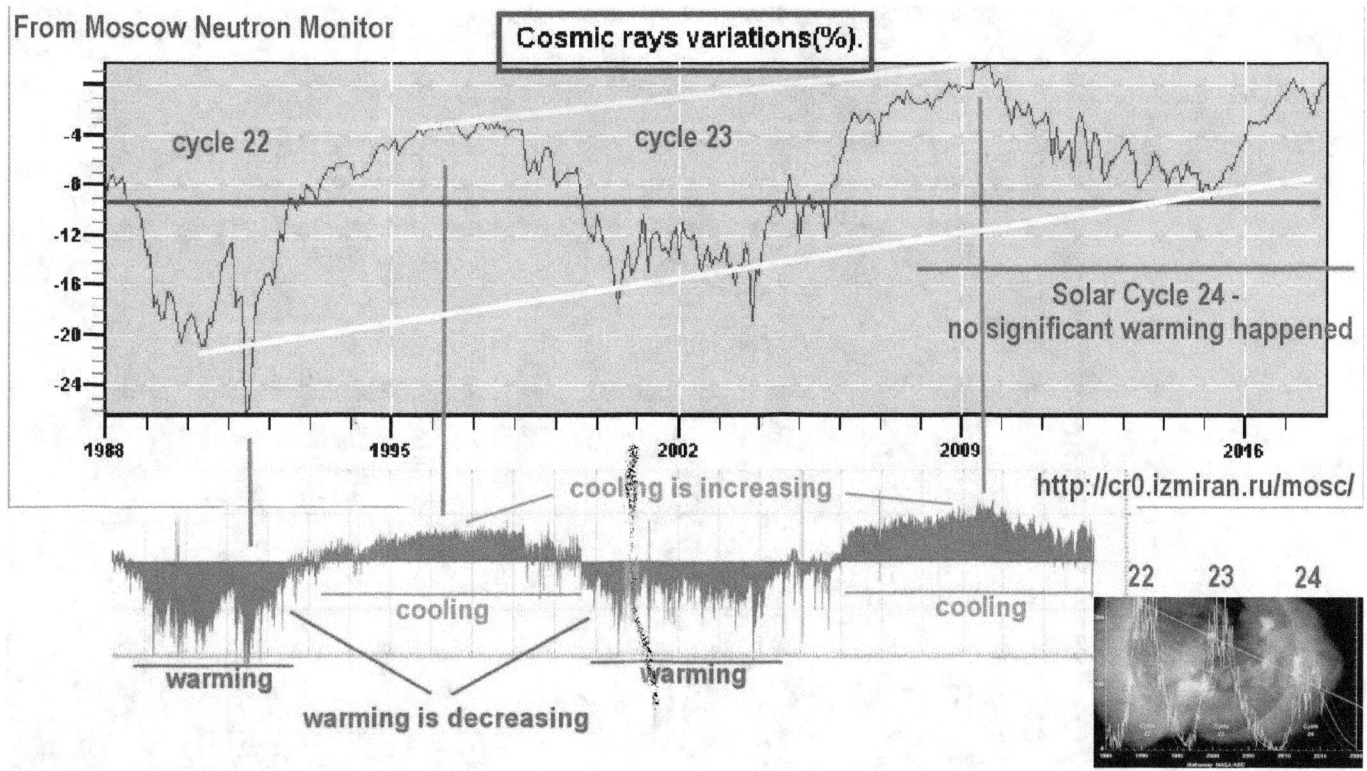

Many of the dynamic features that have been discovered about the solar system, are presently fast loosing ground with the solar system, that includes the plasma streams, getting weaker. We see the solar cosmic-ray volume correspondingly increasing. And that's happening at an amazingly fast rate.

Fringe-effect consequences on the wide field of climate effects

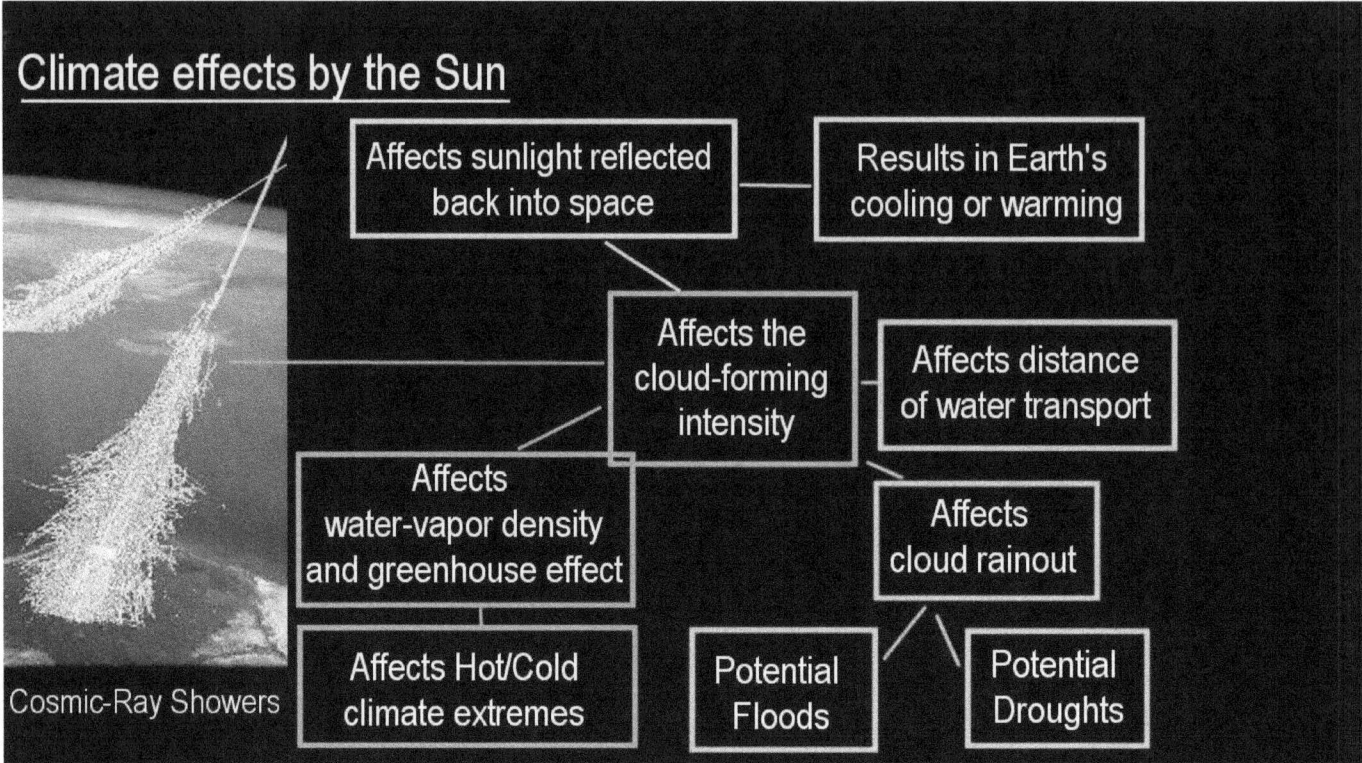

And we see the the cosmic-ray increase reflected in increasing fringe-effect consequences on the wide field of climate effects, from floods to drought to cold and snow.

The greenhouse effect of the atmosphere getting weaker

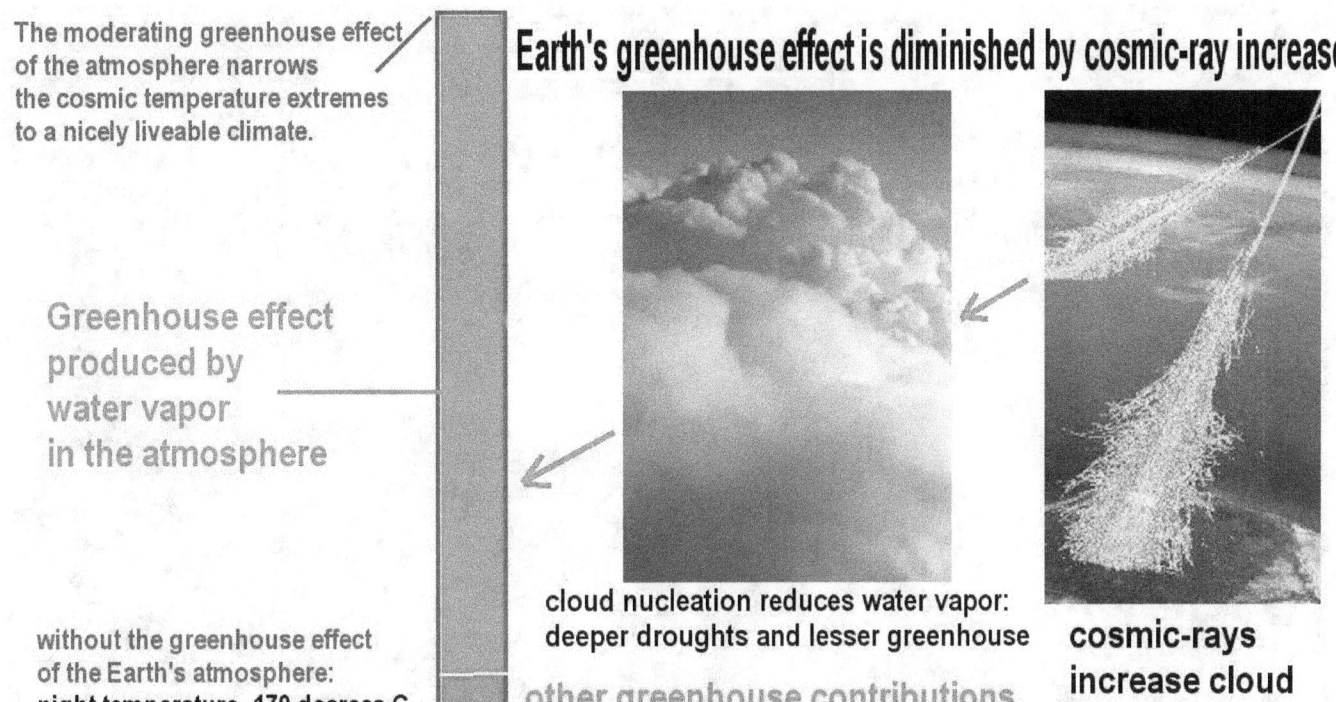

We also see it reflected in the greenhouse effect of the atmosphere getting weaker. And that's happening fast.

Ulysses saw a 20% increase of cosmic-ray flux per decade

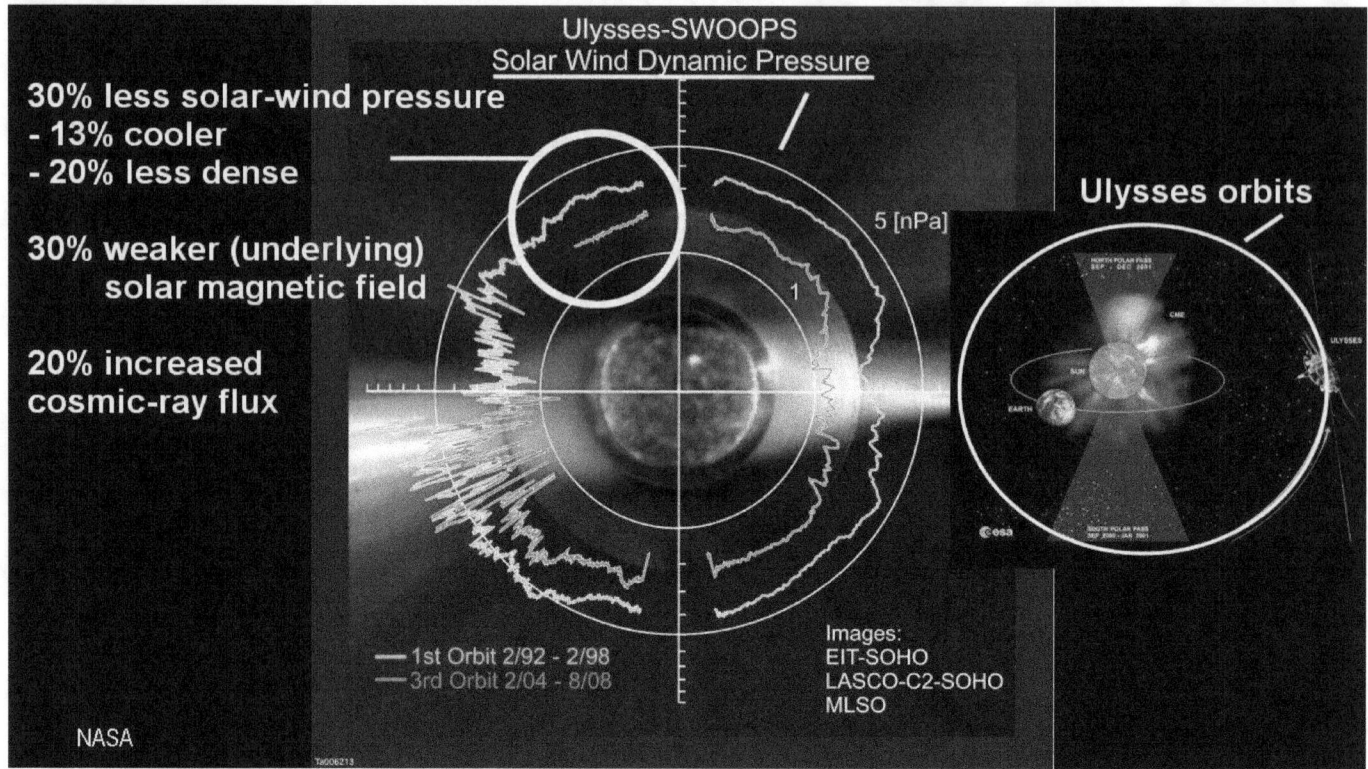

The Ulysses spacecraft saw a 20% increase of cosmic-ray flux per decade over the timeframe of its mission.

We now see the same fast rate continuing

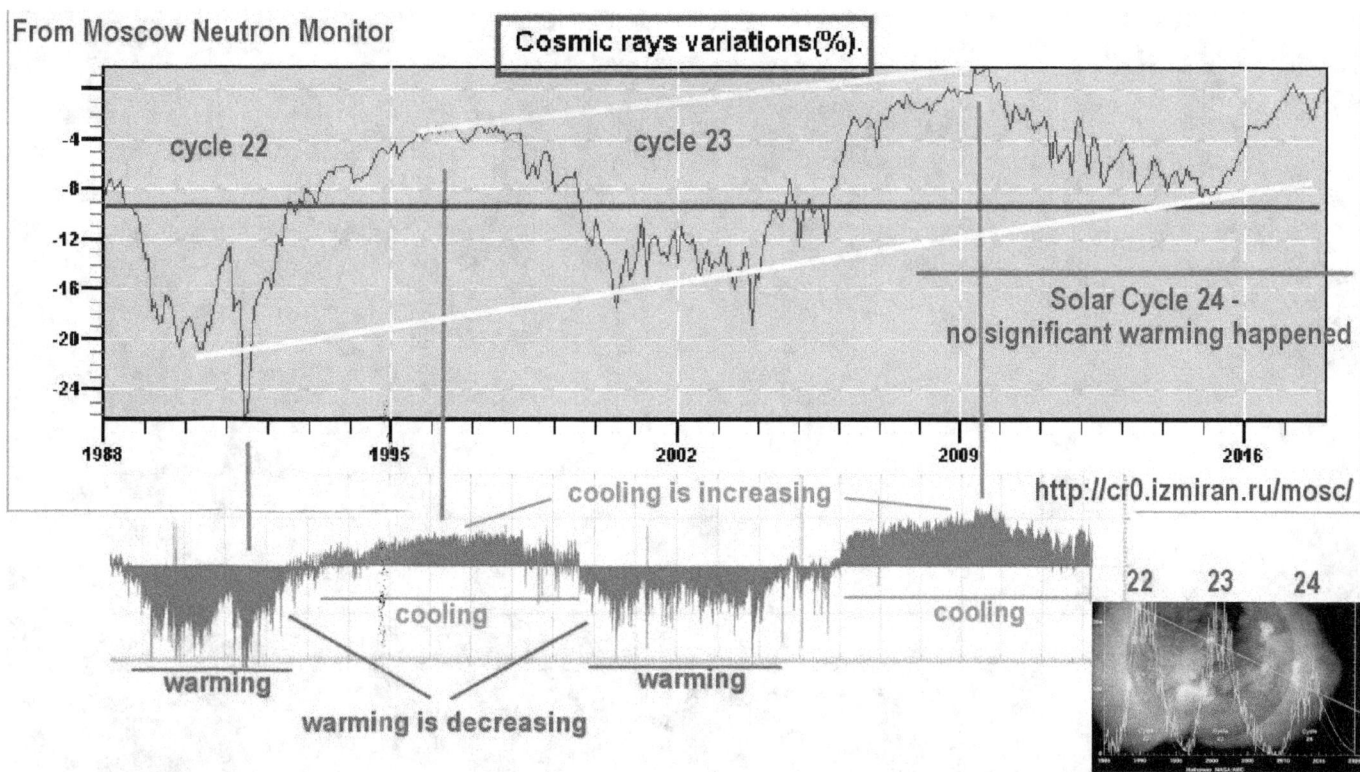

We now see the same fast rate continuing in the measurements of the Moscow Neutron Monitor.

Even the heart beat of the solar system is slowing down

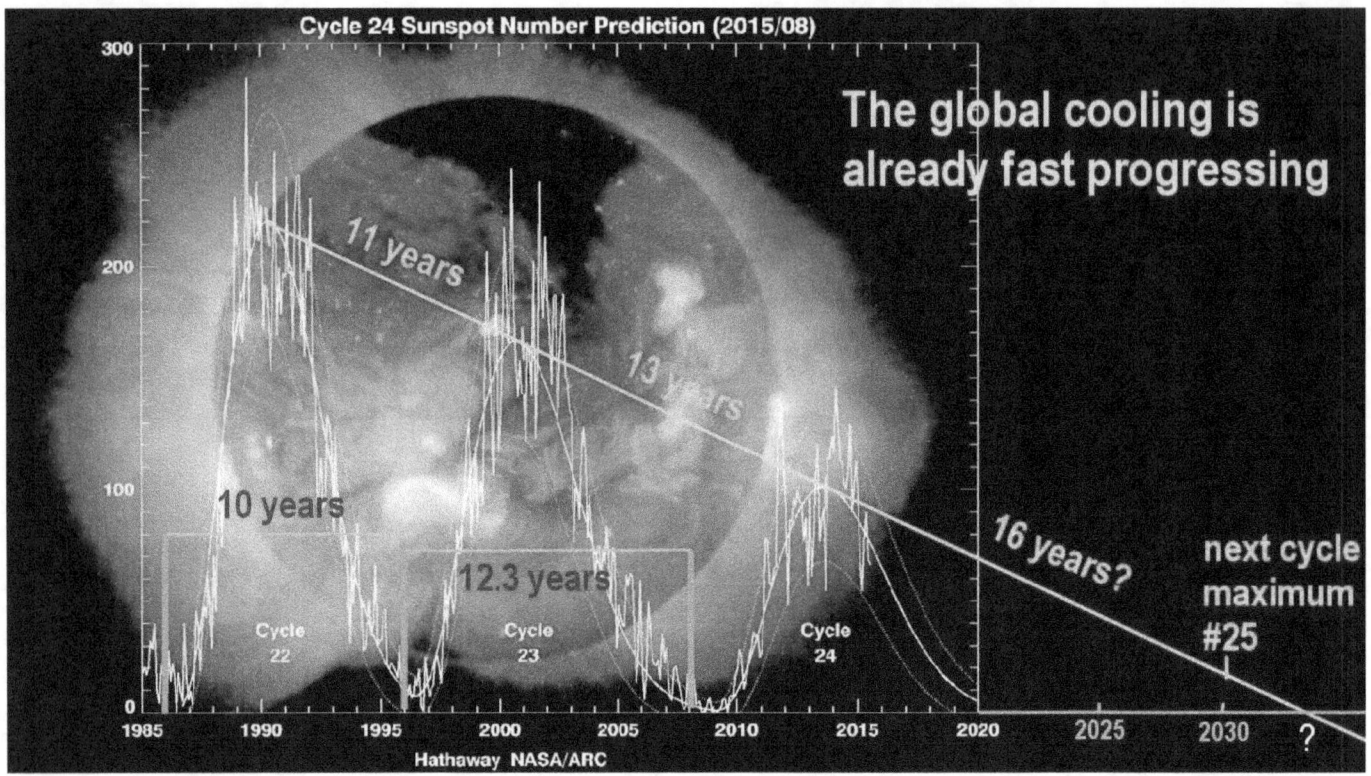

Even the heart beat of the solar system, the 11-year solar cycle, is slowing down and getting fainter.

Real-time measurements become important

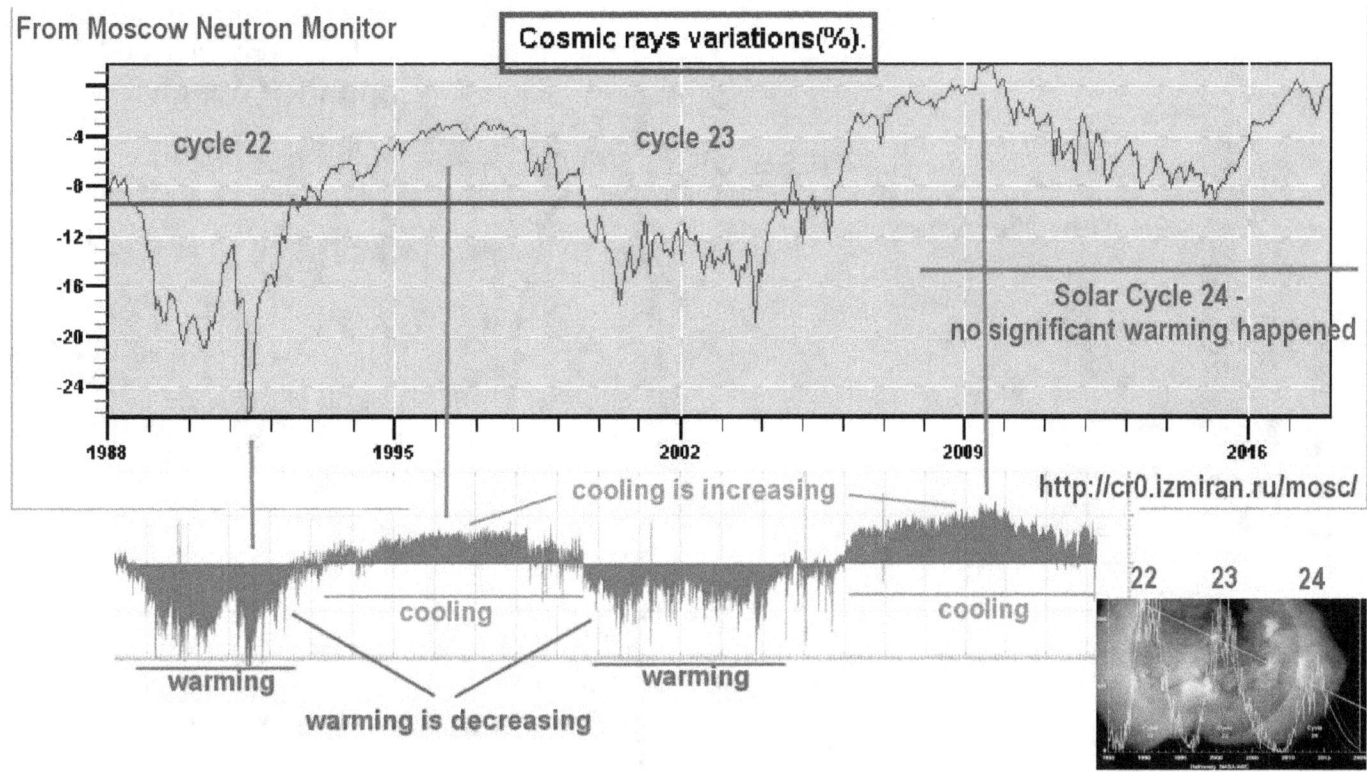

Here, the real-time measurements of our plasma Sun become important, because they speak of a climate collapse in progress, progressing towards the big phase shift into glaciation.

When the primer fields collapse

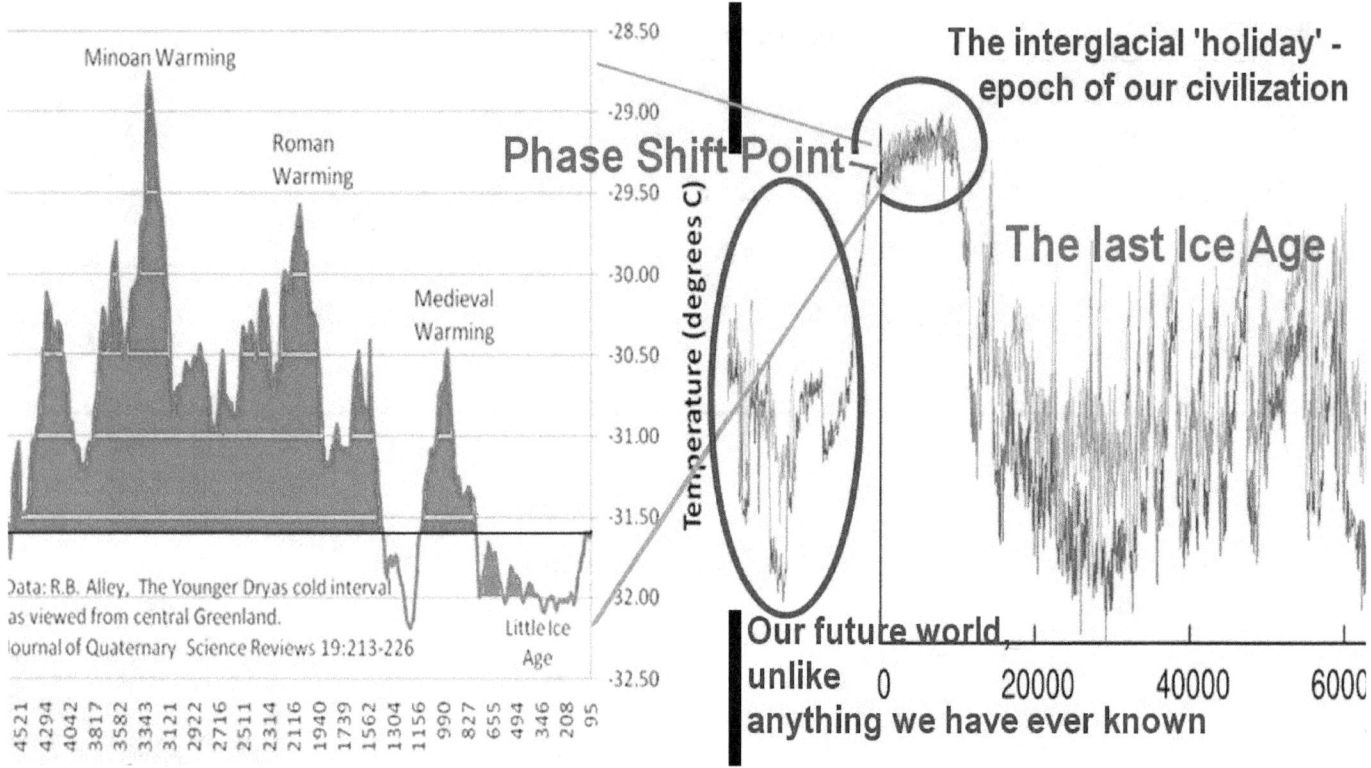

When the primer fields collapse and the Sun goes into hibernation, we find ourselves in a different world with 70% less solar energy being radiated from the hibernating plasma Sun, and 80% less precipitation happening. Agriculture will then collapse, and without agriculture, humanity will vanish.

Our understanding of the dynamics of plasma Sun and the principles that operate it, should cause humanity unite and join hands to build itself a technological new world that the collapsing climate cannot affect. But nothing is done on this front, while the remaining time to get this done is getting shorter.

Humanity has put itself effectively asleep with inconsequential pursuits of wealth and power and war. The strategic defense of humanity is off the table, and time is running out. What will it take for humanity to awake?

We may be only a decade away from the stage when agriculture begins to collapse under the pressure of the fringe effects that are already happening and are increasing, especially in the northern regions. There, we may see entire nations functionally collapsing and ceasing to exist, ranging from Canada in the West, to Europe and to Russia in the East. That's the face of the boundary zone to the next Ice Age unfolding, with processes that cannot be stopped, but for which the consequences can be avoided when humanity wakes from its slumber.

The real, deep Ice Age Challenge stands before us all

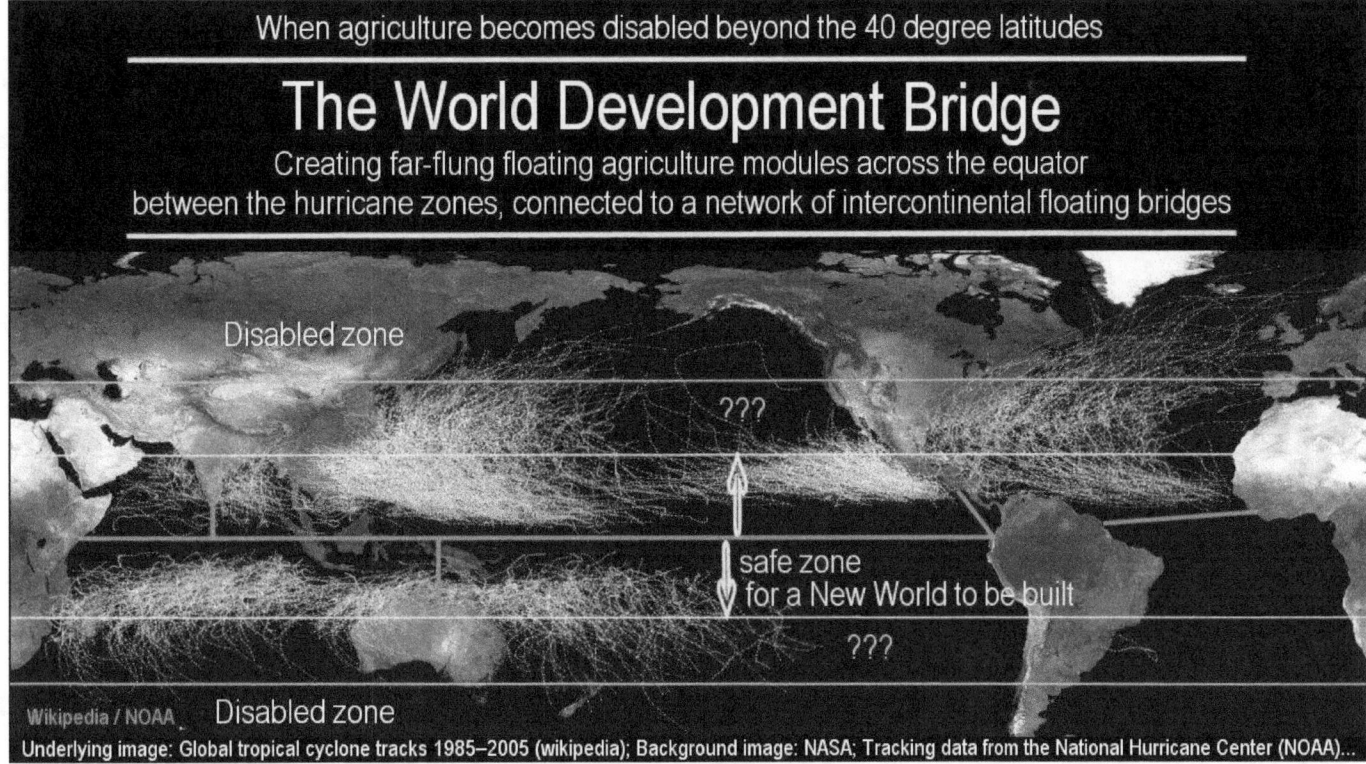

In order to avoid the consequences of the resulting climate change, thousands of new cities need to be built, up to 6,000 of them for a million people each, all with new agricultures and industries attached, and so forth, and with new freshwater resources and new energy resources.

Since little suitable land is available in the tropics, for this global development of a new world, most of the new world will need to be built afloat onto the equatorial seas. Tragically, nothing of the sort is done.

Here, the greatest potential tragedy in the history of civilization is located, as absolutely nothing is being built to face the looming existential crisis, or is even being considered.

That's the real, deep Ice Age Challenge, which stands before us all. It unmasks for us the existence of an ice age within us.

Whether humanity will survive the astrophysical Ice Age challenge that it is reluctant to face, which is greater than any challenge in its entire history, remains yet to be seen.

The astrophysical challenge can be faced, and a new world can be created to avoid the cosmic consequences that only few would otherwise survive, but again I must ask, will it be done? Will humanity create itself the greatest new renaissance ever - the brightest ever experienced - with which to master the astrophysical challenge?

The materials, energy, and technologies to master the challenge do all readily exist. Humanity lacks nothing to build itself a new future, should it raise itself out of its pit of small-minded thinking, and this in time to rescue itself.

The potential for a bright human future is immensely great, in-spite of the Ice Age that cannot be prevented by any means available. But will society rouse itself out of its present smallness at heart and realize its potential?

This question places the most critical challenge of our time out of the realm of cosmic dynamics, and puts it into the realm of society's self-healing of its smallness, so that it will not lay itself down and commit collective suicide by default?

Here is where the scientific recognition of the dynamics of the universe becomes important, including that of the plasma Sun that is a pivotal element of it.

The needed self-healing of humanity is possible

The needed self-healing of humanity is possible, and will likely be achieved when the Ice Age Challenge with all its vast dimensions is not seen as a barrier, but is seen as an invitation to the greatest renaissance in all respects that lies within reach for the grasping.

More Illustrated Science Books by Rolf A. F. Witzsche

www.ingramcontent.com/pod-product-compliance
Lightning Source LLC
Chambersburg PA
CBHW081002170526
45158CB00010B/2879